RailMODEL Digest

Issue No. 4

RailMODEL Digest
is edited by
Iain Rice & Mike Peascod
and published quarterly by RailModel,
an imprint of Hawkshill Publishing.
RailMODEL Digest is sold on
subscription and through
the book and model trades.

RailMODEL Digest is printed in
Cornwall by: Penwell Ltd.,
Station Road, Kelly Bray, Callington,
Cornwall, PL17 8ER

Advertising
Please send advertising enquiries to the
editorial office. A rate card is available.
Space is very limited.

RailMODEL Digest
PO Box 2, CHAGFORD,
Devon TQ13 8TZ
Phone/fax: 01647 433611

Copyright RailMODEL/Hawkshill
Publishing. All rights reserved.

No part of this publication may be reproduced or transmitted in any form by any means, including photocopying or electronic media, without written permission of the copyright holder. Multiple copying of this publication or its contents prohibited without prior written permission.

The views expressed in articles published in **RailMODEL Digest** are not necessarily those of the editors or publishers. Contributions are accepted and published in good faith, but no liability can attach to the publishers for loss or inconvenience suffered by any party due to error or omission.

All reasonable care is taken with advertisements carried, but the publishers cannot be held liable for the actions of advertisers.

Please contact the editorial office before submitting unsolicited material, for which no liability can be accepted.

Contents

The Parson's Charlie 4
The Rev. Louis Baycock builds a Q1 in EM.

Tin Tabs, Ebeneezers and the Piggery Mission 11
Tim Shackleton goes nonconformist.

Arcadia: Colonel Stephens in 7mm. 18
Martin Brent outdoes the Colonel in Kent...

Modelling in Card: Rice amid the cornflake boxes 29

Aye to Eye: Eye-care for modellers 34

4mm. Wagon Running Gear Sprung Systems 38

Prototype Study 42
Hughes L & Y 4-6-0's - The Final Twenty.

Prototype Matters: Cattle Wagons 51

Under the Lens: Some 3mm. Models 52

Pre-Group Glory: The LNWR at Bushey 54

Blagdon, S & D J R 55
A highly plausible might-have-been in 4mm. scale.

Wiring for the Electrically Illiterate 63
Brent on control panels

Carriage & Wagon Dept: Denaby PO wagons 66

MPD: Finishing off the 2251 70

Workshop: The DG Coupling 73

Fox Talbot Tales: Let there be daylight 76

Layout Design: After Garliestown 80

Zap! PCB's, Magnetic Reed Switches 84

S & T: A couple of unusual prototypes 87

T.P.O. Your letters 88

Reviews: HO Hunslet, Flux, Gearbox, Wagon book 90

Front Cover: Autumn comes to the clay country; 57XX and clay wagons in a highly-characteristic Cornish setting. All trickery ...

Modelling and photo: Iain Rice.

Rear Cover: Arrival at Arcadia: Hunslet 0-6-0ST of the Isle of Oxney and Tilling Tramway running into 'Arcadia', Martin Brent's new 7mm. layout featured in this issue.

Photo: Iain Rice.

From the GENERAL MANAGER'S OFFICE

This issue of **RailMODEL** *Digest* marks our first birthday - exactly a year since the Preview issue was launched upon an unsuspecting public at the 1995 Watford Finescale Extravaganza. To those many of you who have supported us so staunchly over that difficult period - for the first year of life of any new-born enterprise is always a difficult period - our heartfelt thanks. To those few who took great delight in predicting our imminent and speedy demise, on the other hand, we respectfully offer a truly Churchillian salute!

We have had to overcome quite a few difficulties in the course of the year - finding suitable storage premises, building a workable relationship with our Cornish printers (it's the language barrier, you know...), sorting out a despatch system able to cope with the trade orders and a subscription list three times larger than we anticipated, battling with the odd piece of re-calcitrant computer hardware, preserving the springs of the much-abused editorial Volvo (a couple of dozen boxfulls of *Digest* at 18 kilos a go weighs, er, rather a lot) and coping with the loss, pro tem, of our invaluable Girl Thursday, the estimable Helen.

Some of these problems are with us still, but things are gradually settling down, and we're now able to really plan ahead with confidence both for the *Digest* and for the modelling handbooks. Martin Brent's comprehensive book on 7mm. fine scale modelling is now entering the production phase, and we're in the process of launching a range of pithy, economical 48-page single-topic volumes with a look at PCB Trackwork, courtesy of Rice who is now building the stuff in his sleep...

From Issue 5, the Digest format will be refined a little - no drastic changes, but a bit of fine-tuning and some more decorative artwork for the various practical departments which are coming together at the rear of the magazine. We have in hand several further tutorial series that will build, issue by issue, into a wide-ranging and highly practical manual of railway modelling, from the obvious to the obscure and right across the scales. Our overseas links are now bringing dividends in the form of articles on modelling techniques, subjects and materials not familiar in this country. Upcoming topics include such diversities as rock-modelling and military railways.

Of course, we couldn't achieve any of this without the efforts and support of our contributors across the world. A huge thank-you to all, and especially to those regulars we depend on issue-by-issue: Graham Baseden - who processes, prints and freely criticises much of our photography - Louis Baycock, Martin Brent, John Chambers, Dick Ganderton, Tim Shackleton and Graham Warburton. All have performed service above any reasonable call of duty.

We also value our international links very highly, and we're pleased that we are now in contact with modellers throughout Europe, the USA and Canada, Scandinavia, Australia and New Zealand and the Far East. The *Digest* is now read in places as far-flung as Indonesia and Iceland, while a substantial proportion of our subscribers come from the Netherlands, where our direct overseas links are strongest. We like to think that this broad web brings a little breadth to our outlook.

Over the year, we also hope that you'll have noticed a steady improvement in our typography and design. We started out with the 'L' plates very firmly on display, but we feel that, with experience, better software and with our AppleMacs bulging with bolt-on memory, we're coping a bit better; perhaps we might even aspire to some of those green 'Provisional' plates for the newly-passed.

Errors, Omissions and All That

But before we get too complacent, it's time to put our hands up to a boo-boo in the last issue.

We need to credit a photograph used in Graham Warburton's piece on 'Long Tom', the extraordinary LNWR double-bracket signal, in Issue 3. The prototype shot at the top of p.56 was taken by Graham Turner. Sorry about that.

Photo credits are a bit of a problem for railway publishers generally, with thousands upon thousands of pictures in free circulation - many totally bereft of any identification as to photographer or origin. We credit where we can, but it's obviously impossible to establish a pedigree for every picture. If there's a problem, we're always happy to put right any errors and stump up a few quid as a repro fee.

Iain Rice and Mike Peascod

From a SHED on DARTMOOR

Wot, No sky?

These days, hawking the old Hawkshill stall around the shows as well as dashing hither and thither about the country with the cameras means that I 'm getting to see rather more model railways than would normally swim within my ken down here in deepest Dartmoor. And, more and more, I'm struck by what an immense difference the proper use of a backdrop makes to any model railway.

Since No. 1 daughter started at Big School last September, I have added an early-morning drive across the moor to Ashburton to my daily routine. Which, at this time of year, is no hardship and can be a positive joy; for my money, England in general - and Dartmoor in particular - is at its rich and varied best in the early Autumn. But what I have noted, in observing this tract of countryside on a daily basis, is just how influential the changing aspects of the sky are on the whole look and feel of the landscape. The sky is, in so many ways, the single most visible and critical element of any landscape.

This is true enough in hilly terrain like Dartmoor; but skyscapes are an even more prominent factor in low-lying countryside. It was something of a preoccupation when I was trying to model East Anglia, and now that I am tussling with the modelling challenges of the unique (but undeniably *very* flat!) Dutch landscape, I find that the sky is once again looming large in my list of problems and priorities. To me, much the great charm and appeal of the Netherlands lies in the combination of that vast expanse of sky, held in glittering counterpoint by the myriad watery elements of the landscape.

On a model, I don't see how you can possibly recreate this feel - the sky, and its reflection in water - if you haven't *got* a sky in the first place! To a very large extent, the sky *is* the model, and I certainly can't conceive of an effective way of presenting the landscape without that other, dominant half of the observed scene. And there's no way that one could, in my view, create the illusion of Holland (or East Anglia, or East Cheam if it comes to it) against a backdrop, not of the luminous blues, whites, creams and greys that make up a typical skyscape, but of totally discordant and irrelevant elements: jazzy jumpers, verbose T-shirts (or, in Martin Brent's case, best shirting and snazzy bow-tie) - elements that have more to do with the sartorial sense (or lack of it) of exhibition layout operators. Many home layouts aren't in much better case, either; the rose-ranked distractions of designer wallpapers have more to do with the interior design aspirations of the modeller's better half than with the natural presentation of the model!

It is notable how virtually all model railway publications - this one included - supply (or attempt to supply) some sort of sky background when photographing layouts, even where the layout itself is bereft of any sort of backdrop. This may, indeed, be the reason that I am so often disappointed when viewing in the flesh some layout only previously appreciated in magazine photographs. The influence of that supplied sky, its neutrality and the enormous enhancement that even the plainest of backgrounds can bring to the appreciation - without distraction - of what has actually been modelled is almost immeasurable.

Why, then, are backdrops so neglected? It does seem ridiculous, when you appreciate the care lavished on the modelled aspects of so many layouts, that all this is so largely negated by allowing it to be invaded by discordant and irrelevant visual clutter from behind, while blithely ignoring so significant a segment of reality.

Well there are, of course, practical difficulties in installing backdrops, particularly on exhibition layouts; but they are not insuperable, as modellers like Barry Norman and Vincent de Bode have showed. Martin Brent's 'Arcadia', our main layout feature in this issue, demonstrates that even a very simply-executed backscene expands the visual boundaries of what is actually quite a small layout enormously, and helps to endow this delightful model with that wonderful 'garden of England' character. The practical problems of layout operation from behind this sweep of hardboard Kentish sky were solved with a simple step-up box for the operators, a device also used by Norman and de Bode; the alternative is the front operation favoured by the Ulpha Light crew - see Digest 3 - not to mention yours truly.

Considering the backdrop as an integral part of the layout from the outset has a number of advantages. For a start, it can become an important structural element in the baseboards, giving great vertical stiffness (ie., lack of 'sag'); it can also open up possibilities for representing a great deal more on the layout than can be fully modelled in the space available. I'm more than ever convinced that the creative use of a backdrop is a fundamental element in the creation of an effective model railway; let's have more of 'em.

Iain Rice

The Parson's CHARLIE

4mm./EM

O. V. S. Bulleid was one of the most controversial locomotive engineers of this century, but nothing he proposed was more startling, at least in terms of appearance, than his stark but highly practical 'austerity' 0-6-0 goods loco, the Q1. Following the Continental system, Bulleid classified these 0-6-0's as a 'C' wheel arrangement (3 powered axles) - which, under the phonetic alphabet code widely in vogue at the time, at once became a 'Charlie'.

There are two kits for this ubiquitous engine currently available in 4mm. scale - etched brass by Crownline, or cast whitemetal by Finecast. It was the latter that the **Rev. Louis Baycock** selected as the starting point for his EM model.

Introduction

Once again nostalgia has me in its grip and since I have a modelling interest in exchange traffic in and around London, the Southern Q1s I remember so well are, for me, a 'must'. One could often be seen at Harringay waiting to go down to the Tilbury line, and having admired the beast as rugged and workmanlike, even handsome in a 'macho' sort of way, when Finecast brought out their revised version of the 'Little Engines' whitemetal kit I bought one 'for stock'. In the event, I found it interesting and instructive as a first essay in white metal kit building, learning the hard way, as will be seen!

The Finecast Kit

As with all South-Eastern Finecast kits, the presentation is not only attractive but useful - a strong box capable of storing the finished model, parts well-packed with lots of tissue paper, plenty of instructions and everything needed bar the usual motor, gears, wheels and finishing materials; all very satisfactory and helpful. However, one drawback soon manifested itself: the instruction sheets needed copying and sorting in proper order. This may seem a minor point, but I found it much better to have them in numerical order - to turn over page 1 and find oneself on page 4 is a bit disconcerting! On the plus side, the parts list is very helpful, and the little bit of history/livery information plus the detail construction notes were all of interest - although these latter should be read in conjunction with the general instructions, which I fancy I recognise as some of our Editor's handiwork from earlier days. Diagrams are provided for the fixing of pickups, and the double-sided exploded drawing is lucid - although if one is not used to engineering drawings, it takes a bit of 'seeing', but light dawns eventually!

All the bits came well protected and listed, with thoughtful provision of duplicates in some cases (I found them very necessary!) The castings were very good, with a minimum of flash and part-lines. The dome and chimney did need some attention - the top of the dome overhangs quite a lot, and my specimen was a bit tight to the sides. The chimney rim also repaid some slimming down, for which I used a file tang - an occupation to be carried out patiently and gently when in a calm and reflective mood.

The etched parts are on a well-designed fret with minimal attaching nibs, the parts requiring little tidying up after removal. (Why *do* so many kit designers find it necessary to smother every component with umpteen joining tags? It makes for a lot of work and a far greater chance of damaging the etched components.) All-in-all, a most satisfactory kit.

Components to Complete

Some very nice wheel overlays are provided convert standard Romford wheels to the Q1's distinctive Bullied-Firth-Brown boxpok-type, but luckily for me the truly excellent Markits/Romford wheels arrived on the market at just the right moment; these have the correct domed appearance, are available in the finer (and, for EM, preferable) RP25 profile - and fit Romford axles, so quartering is simple (Oh bliss! Oh joy!). They look, to me, just right. To complete the mechanical specification, I chose to use a Mashima 1628 'can' motor with flywheel driving through a Branchlines 2-stage gearbox with an overall reduction of some 53:1 - good for slow running.

The prototype Q1 (Charlie) in all its stark glory. There is an apocryphal story that, when No. C1 was unveiled at Waterloo in 1942, Stanier - who was present as a guest - leaned across and said to Bulleid (soto voce, but not too soto, according to the account) 'Tell me, Oliver; where do you put the key in?' (Source: The late C. R. H. Simpson, sometime editor of 'Engineering'.)

Tools, Preparation

I am not the possessor of a particularly extensive tool kit, so it's just as well that the Finecast kit makes modest demands in this direction. A Stanley knife, a small selection of files, a pair of Antex instrument soldering irons (so as to avoid mixing lowmelt and ordinary solder on the same bit, never a good idea), tweezers, snipe and flat-nose pliers, small snips, a jeweller's screwdriver and some small drills (0.45mm., 0.7mm., 1mm. and 1.5mm) plus taper brooches and a $^1/_8$ in. parallel reamer were the basic requirement. Also useful throughout the making of this kit were 'Body Shop' large emery boards (in lurid colours, having coarse and fine sides) and very fine (320) grit 'wet & dry' abrasive paper.

I found it invaluable to go through the whole kit tidying up parts, mostly with a fine file used without too much zeal. Trial assembly of castings with 'Blu-Tack' was very helpful to check the fit of parts. Time taken at this stage is time well spent. I found that when drilling holes, particularly in whitemetal, it is better to drill small then to take time over reaming out with the taper brooches, thus obtaining a snug fit for wires etc. Plenty of wire is provided in the kit but it is cut from coiled stock, so I used a combination of this and Alan Gibson's straight hard brass wire, which proved rather better for some of the straight pipe runs.

Assembly

I chose to solder my model together. Throughout the assembly process I used two soldering irons, one for the 145° solder (brass/nickel and tinning ready for whitemetal solder) and the other exclusively for whitemetal. This has had a 'point' bit supplied by RS, which in fact has a small ball end; kept clean and hot, it proved ideal for 'stroking' seams and 'pointing' joints - always with liberal applications of flux.

The principles of kit-assembly soldering are enunciated in editor Rice's 'Whitemetal Locos - A Kitbuilder's Guide' (Wild Swan Publications) and I found they worked very well; in general I used liquid phosphoric acid flux as advised, but I also tried 'NSF' flux paste, made in the USA by LA-CO Industries. Inc. of Chicago, which carries this description: 'Cleans as it solders. Non-toxic, non-acid, lead free'. It proved very effective, particularly when soldering handrails - it does not run everywhere before you can get to the spot with the iron! *(Eileen's Emporium now sell a very similar - and highly effective - paste flux - see the review notice in this issue - IAR)*

Although the kit instructions say solder or epoxy may be used to join components, I have found the latter unsatisfactory in the main, preferring to solder - even whitemetal to brass, achieved by tinning the brass with 145° solder and then fluxing well so that the lowmelt solder would run between the faces by capillary action, which it did very well. One great advantage in using lowmelt solder is that if you get it wrong, a swift dunk in boiling water undoes your handiwork, enabling a fresh start to be made. I even used this dodge selectively for one component, without 'getting a kit all over again'; probably beginner's luck!

Chassis Construction

When starting work on the chassis, I obeyed the instructions and first drilled/opened out all the necessary holes. This chassis is not really designed for compensation, which I wanted to fit, so this work included careful drilling of the compensation beam pivot holes centrally between the front two driving axles.

I found it easier to punch out the half-etched rivets in the frames after parting these from the fret. I used a blunt scriber and a gentle tap or two with a light hammer to emboss the rivet-heads. However, prototype photos show that these rivets (or are they bolts?) at the front end stand out quite noticeably, and I did not make mine 'heavy' enough, and unfortunately it was too late to improve on them when I realised their lack of sufficient prominence.

While the frames were still separate, I cut out the front hornblock openings (half-etched guides provided) and trimmed the bottom leading corners of the firebox sides to clear the actual hornblocks, which in this case were the MJT 'simple' fabricated type. At this stage, I also carefully fitted the bearings of the rear (fixed) driven axle to the frames and tacked them in place.

The frames could then be soldered to the frame-spacers. I model in EM, so Alan Gibson's L-section EM spacers were made up and used strategically. When not using the Finecast spacers, it was important to make sure that the holes in the horizontal part of the replacement etched spacers came in the same locations as those in the turned brass spacers to facilitate attachment of the body.

I also found it necessary to use a *small* etched spacer at the front of the chassis, to leave room for the cylinder block to come above it - it could take some 'ackling' to get a fit here right later! Apart from this, I fitted two flat spacers at the top and bottom of chassis between the two large lightening holes in the frames below the firebox to provide motor and collector Veroboard mountings. Careful positioning of spacers, pick-up mountings and other between-the-frames components was needed as the firebox is the only thing that should show through these weight-reduction holes in the frames.

If building a non-compensated 16.5mm. gauge model, following the instructions would produce a satisfactory rigid chassis with the sides bolted to the turned spacers provided, although I would prefer to fit extra spacers soldered in to hold everything firmly in place - where I'm concerned, screws seem to have a habit of coming loose.

Motorisation
For DS 10 or RG4C motors (as recommended) use the motor mount provided and work out your own salvation with the instructions. I used a Branchlines Multibox with my chosen Mashima 1628 - which proved a good combination, easy to assemble and accurate in mesh, with consequent smoothness of running. The original Multibox design has now been improved by the addition of 'ears' making the mounting screws accessible so that the motor can be removed without dismantling the gears, which I found a great bonus.

Using this setup enabled the motor and flywheel to sit in the boiler - after some surgery on the castings, of which more anon. The gears selected gave a reduction ratio of 53:1 with 2mm. motor shaft; the bottom holes in the etch were used, and with everything meshed, lubricated and tested, I installed it in the frames and did some running in, fettling the frames of the gearbox to get the smoothest running. That done, the gearbox was removed from the chassis, the redundant parts snipped away and the edges rounded off, making a neat job that was quite unobtrusive and cleared the body castings. It really all went together very well - my best effort yet!

I also found that lubricating this geartrain with a very small amount of a heavier oil (I used nice tacky chain saw oil) helped to keep things quiet and the lubricant in place, rather than oil being centrifuged everywhere as often seems to happen with normal 'model' oils. But I found (the hard way!) that it must only be a very small amount if the most awful mess was not to result.

To give some degree of sound-deadening, I mount my motors to the chassis with Velcro sticky-back pads; in this case a small square fixed on the upper chassis spacer, with a matching piece on the motor body; this gives a little float but enables ready detachment - the motor can be easily removed without disturbing the rear axle and attendant gearing. The rear axle and gearbox were located in the chassis, with thrust and packing washers to eliminate side play; the Romford wheels could then be added at the appropriate time.

The drivetrain used for my version of the Q1. The Mk. 2 Branchlines Multibox proved a good choice, easy to use and accurate, with good gear meshing. Combined with the powerful, slow-revving Mashima 1628, it gives excellent slow running and plenty of haulage power - just like a real Q1. The flywheel is probably a bit of a joke - but what else do you do with all that shaft sticking out the back of the motor?

The Finecast Q1 chassis as adapted for compensation with MJT hornblocks - note the provision for cutting out the hornguide openings on the rear axle, left fixed on my model. As shown here, the wheels, motor and gearbox have been removed ready for the chassis to be painted, which is why all the sandboxes and brake gear have been fitted.

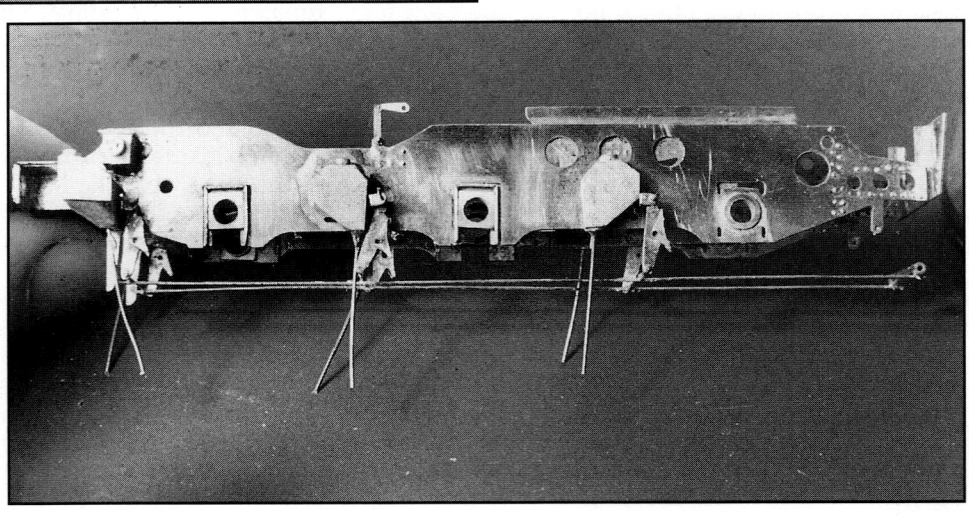

Compensation

MJT Flexichas hornblocks were used for the compensation setup, following the instructions that come with them. For installation, alignment jigs are a must - I have a set of aluminium jig-axles coned at each , supplied by London Road Models and these - with springs to hold everything in place - worked beautifully. The hornblocks were set in place using the coupling rods as location; this was done before the rods were modified, as described in a line or two. When soldering hornblocks to frames, I find it pays to use a minimum of flux, otherwise 'hornblocks, they no slide'!

Coupling Rods

The coupling rods are fine as supplied for rigid suspension; however, for compensation, some thought and careful cutting is needed. I did one side at a time, removing only the parts needed for that side, since it avoids confusion (!). Cut the outer (fluted) etch between the centre crankpin hole and the knuckle, and the inner half ahead of the crank pin hole; they can then be made to overlap at the centre crankpin.

Having tinned the joining faces of the etches I lined them up and held them down with a heatproof implement (half a clothes peg!) and sweated them together with a minimum of flux. The finished rods were then cleaned up with a fibreglass brush, and the crankpin holes carefully reamed out to a slightly sloppy fit (about 5thou. fore-and-aft play) on the crankpins.

The instructions suggest that the chassis should be painted at this stage - I did so, and regretted it, since more pipework needs adding and must match up with the body in various places. The instructions also show the steam reverser as being attached to the body: this I felt could be damaged when the body is set aside, so it was simply soldered to the LH chassis side in the same position, via a small piece of wire with a 'kick' in it. The actuating crank and connecting rod were then located to match.

Now was the time to screw the Romford crankpins into the wheels, fit and quarter these on the axles and fit coupling rods. After testing for free running, I soldered the turned retaining washers carefully to the crankpins, using paper washers to stop the solder running and to give clearance. The Romford crankpins and washers may not come with the Markits wheels, but are on sale in most model shops.

Pickups

The current collectors were now fitted, a small piece of Veroboard PCB being epoxied to the lower central frame spacer, and the wipers (Gibson 0.45mm. hard brass wire) soldered on and sprung lightly against the backs of the wheel tyres. As I had also decided to fit tender pickups, I found it best to use some 0.7mm. wire - suitably bent - as leads to the motor tags, to give an easy attachment for the feeds from the tender. I then indulged in some test running, joggling, lubrication and general fettling of the chassis to get smooth running; it was time well spent.

Cylinder Block, front footplate

The chassis being complete, it now needed the cylinder block fitting, together with the piston-valve covers that form so prominent a feature of the Q1. I added the cylinder covers after fitting the block, as they needed filing to fit, and then used epoxy to stick them in place. Locating the block is easy, but care must be taken to get it lined up in the frames crosswise and aligned vertically with the mounting screw hole in the spacer. In OO , the block needs no attention, it just fits the space; for EM, I had to solder spacers onto the sides to make it the necessary push fit between the frames - offcut pieces of chassis fret filled the spaces nicely.

Building to EM frame width and clearances creates some gaps and misalignments with the body parts, and thought is required. I fitted the various major parts as per the kit instructions, and found, for instance, that the front footplating - very visible - was too narrow where it fitted between the frames; I had to back it with a flat frame spacer and fill the gaps at each side of the footplating between it and the frames with solder.

Body Assembly and Modifications

I then gathered together all the boiler, firebox and cab pieces, chimney, dome, etc. and had a 'dry run' of the body, making sure everything fitted snugly; the joining seam along the top of the boiler sides was a bit prominent, so I gently reduced it a trifle. The assembly diagram is quite clear and helpful and, with the written instructions to explain things, I found no problems (but don't forget to solder the body-fixing nuts in place inside the smoke box and cab!)

Quite a few adaptations were needed to get the front end looking right, especially around the front footplate and valve chest areas, where it was necessary to adapt the kit to the EM frame spacing, particularly in backing the valve chest cover and in widening that section of the front footplating that fits between the frames.

With the motor mounted in the boiler, however, the centre body bulkhead - part 37 - needed to be cut away to clear - I did some of this cutting before assembling the body, taking care not to distort the part, and adjusted the fit once everything was in one piece.

I should imagine that using epoxy, body assembly would be a start/stop procedure to allow the various joints to cure, but with solder it was simple, provided that plenty of flux was used. It paid to get the parts together absolutely tight then, working from the inside, I spot-soldered the castings together. I then re-fluxed the joints well, loaded up the iron with 70º lowmelt, and started seaming. The solder fills the gaps between parts very well, but you do need a clean iron and clean parts. Solder should be visible on the outside of the seams - any excess can be removed with a craft knife - carefully!

When I had a nice square boiler, I carefully carved away the rest of bulkhead to clear the motor, which could be done without any loss of rigidity - but I was careful not to take away any more than is really necessary. The assembled boiler was carefully cleaned up with knife, scrapers and fine wet-and-dry. Visible edges of the castings - cabsides, roof and so on - were chamfered to thin their appearance.

Fittings and Details

I used cyano to fix the chimney and dome, safety valves, whistle and cover. The photos may show I used turned brass safety valves on my model; I broke the whitemetal ones off turning the boiler on its top, so if using the castings provided I'd suggest you leave the safety valves off until last!

When fitting some of the body and chassis details (Finecast instruction 10), one or two things need to be done beforehand if, like me, you're desirous of soldering the bits in place. Firstly, I tinned the appropriate places on the frames for the sandboxes and lubricator, to facilitate attachment of these parts with 70º lowmelt, which won't 'take' on bare nickel-silver. The sandboxes were, of course, fitted before the chassis was painted!

On the middle sandboxes, in addition to drilling the holes for the sand pipes, I gently reduced the outer faces and shaped a piece of 10thou. Plastikard to match their width, extending this beyond the bottom of the sandbox so that it ended up square and level with the bottom of the brake hangers; these overlays were then fixed in place with Cyanoacrylate. This was to reproduce a modification found in prototype photos - I am told that it was made because loco crewmen, standing on the coupling rods (strictly forbidden!) to oil the motion, had their toes lopped off when their mate was instructed to ease forward or back, so that another part could be oiled! Still on the sandboxes, the sandpipes - fitted to the boxes before these were soldered in place - were bent to shape after the wheels were in place.

Following the instructions, I had some difficulty understanding the reason for drilling the lubricator; I used the stud on the back instead of rod in a hole as suggested, and found it came in just the right place when soldered to the frame with a packing washer behind it. The cast buffers were also soldered on now, and filed back to give minimum material behind the buffer beam to leave room for the etched front steps - which, along with the other sets of steps, get a special section in the instructions.

At this point the Vine drawing of the class, plus a selection of prototype photos c/o the editor, swam into my ken and various things became rather clearer as to the alignment of pipework etc. and enlightened me as to what was missing. There's a lot of daylight under the footplate and the chassis ends - space which should be filled by the brake actuating gear. The brake cylinder is under the front of the cab footplate immediately below the backhead, and when actuated transmits the force backward to a hefty cross-shaft at the rear of the engine that in turn pulls the brakes on. This requires pieces to be added to the chassis, shaped as in the diagram, cross shafts and bars soldered in, and the brake draw rods soldered on in their turn. This I found very satisfying on the completed model; too often this is left out, as here, and would be a useful addition to the chassis etch - I don't like to be able to see through here - one can't on the prototype. Now to the pipe work - oh boy!

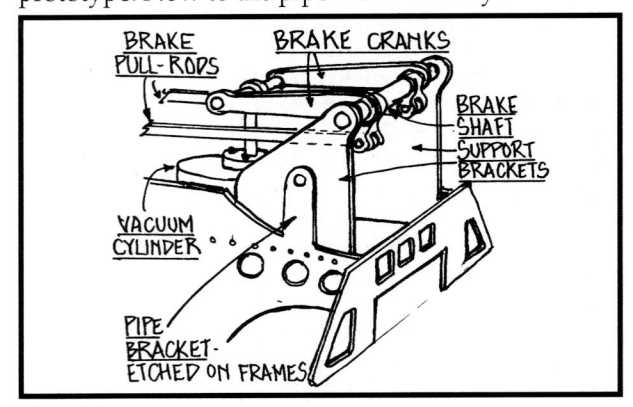

Plumbing

From the detail photos I found, this part of the prototype's anatomy seems to be 'theme and variations', so I just went ahead and did what I saw on one loco, or as near as I could get; so please, purists don't start shooting! Finecast's instruction 7 was helpful, and the drawings likewise, but basically I put boiler and cab to chassis and wheels and worked by eye from there - using various thicknesses of wire plus small washers and fine brass strip - bending and soldering and eventually getting something that looked right.

Most prominent pipework is the combined injectors on the offside; I started from the boiler end and finished at the injector overflows, sliding the cast injectors on while the pipes were straight - the castings provided need splitting and the holes enlarging. The whole lot goes together with the pipes soldered at the boiler end, and with the injectors soldered to their operating rods (Gibson 0.33mm. wires set in the footplate) it is a sturdy piece of work. It did take time and a little touch of 'asbestos fingers'.

At this stage, I also soldered on any pipework that was needed to link to the chassis, fixing it to the frames (having scraped away the carefully-applied paint!) and cutting holes in the boiler base to take the upper ends - it looks messy to stop short, and pipes need to look as if they go somewhere to some purpose!

That finished the loco body and chassis; it runs, needless to say, without any additional weighting, the cast body being quite sufficient.

The model basically complete, with the tender a step ahead of the loco. The seam where the sides of the casing joined the top proved tricky to eliminate, although the prototype also has a 'join' at this point. Note the injector pipework on this side, the fall plate arrangement, and the small number of handrails.

The Tender
The tender was a very satisfying thing to build, and the instructions were simply followed to the letter; the wheels are BFB, and come with axles in the Markit set, but I came to grief in two places by melting the centre bulkhead when soldering the end to a side, and by resting the hot iron on an axle box! These blunders necessitated the boiling water disassembly technique - boil up a saucepan of water and lowering the job into it, removing the pan from the boil immediately; this gives you your kit back! The bulkhead was repaired by laying it smooth side down on plate glass, fluxing well, then placing a shaped piece of the bulkhead removed from the boiler, soldering it in pace, and finishing off with an emery board. It worked!

My other blunder was again negligence, i.e. not looking what the rest of the iron was doing; this time I only dunked the affected part in the boiling water, and happily, only that detached, praise be! A phone call to Finecast who, for the princely sum of £3, cheque with order, supplied a replacement part by return of post: medals all round, except for me!

I added the refinement of tender pickup; in EM there is room on the inner tender chassis sides for a piece of PCB sleepering to be glued behind the wheels, with phosphor bronze strips bearing on the wheelbacks. Holes were drilled through the strip and the chassis and fine wires led to a brass-tube-and-split-pin plug-and-socket arrangement mounted on a further piece of PCB at the front of the tender -see photo below. The connections to the loco look just like water feed pipes, and are soldered to the motor feeds on the loco chassis.

Details and Refinements
Three things I found needed to be added for a bit more realism: a piece of chequer-plate to ride over the loco cab footplate as a fall-plate, an end extension to box off the end of the fire iron tunnel, and a second filler on the tank top, which I made from a 6 mm. square piece of brass with a hole in the centre (I used a cut down frame spacer) into which was soldered a brass bolt with a 4mm. slotted head, filed down and with a flattened piece of soft brass wire in the slot to form the catch.

The buffers on the tender should be made up with the rear steps and then fitted to the buffer beam, as per instructions. I found this a bit tricky. Beware also the long tender ladder, which is very fragile, being I think a little over-etched, but it does look good! Other finishing touches before painting were the addition of vacuum hoses, heating pipes etc., for which I used brass wire with fuse wire bound around and lightly coated in solder - it's no good, I break whitemetal ones! All handrails (not many on a Q1) were made from the wire supplied and force fitted into holes reamed out slightly undersize holes. Lamp brackets were made also from wire, flattened with pliers at one end, the opposite end being soldered into holes drilled to match the positions of the lamps.

Body Mounting, Couplings
Final tips: It proved much easier to detach loco from chassis if only the front bolt was used, suitably shortened, with two pieces of 0.7 or 0.9mm. wire shaped and soldered under the cab to point backwards and fit into the inner corners of the triangular openings in the rear drag plate.

For coupling loco to tender, I found that the specified method would not work easily with the added brake gear on the loco, so I soldered a small piece of brass tube centrally on the rear drag plate with a piece of 0.7mm. wire pushed to swivel in it, and bent in a 'square S' shape with a loop on the outer end to go over the post under the tender; the inner end needs piece of 0.33mm. wire soldered at right angles across it, to prevent too much swivelling: it all works well. At the front I fitted a Jackson screw coupling, but for actual use will be fitting the Alex Jackson pattern.

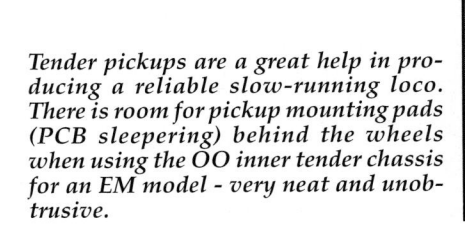

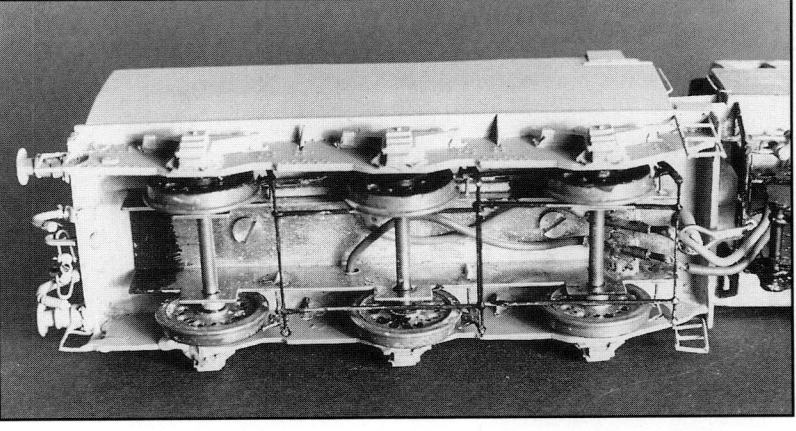

Tender pickups are a great help in producing a reliable slow-running loco. There is room for pickup mounting pads (PCB sleepering) behind the wheels when using the OO inner tender chassis for an EM model - very neat and unobtrusive.

Painting

After washing everything (sans motor!) in weak washing soda solution it was dried with a hair dryer and greasy fingers kept off thenceforth. Before painting I washed it all in meths. to make absolutely sure of a grease-free surface. Painting itself was straightforward: the chassis and wheels were brushed with Humbrol acrylic satin finish black, which in my view gives a slightly 'oily' look, and the body and tender aerosol-sprayed, using a hair dryer to overcome any damp and to dry off the volatiles quickly. A coat of grey acrylic primer was followed by a coat of Humbrol 'Krylon' Ultra Matt Black, which makes a very good basis from which to weather suitably. Number and logo transfers were Methfix, and the final job was weathered by editor Rice, using drybrushed acrylic and Carr's weathering powders

Conclusions

All in all a very stimulating and satisfying kit to build, that goes together well and looks right when complete, especially with the Markits wheels. If asked, I'd give the kit, as marks out of 10: Presentation: 10; Instructions: 7; Quality of parts: 9; Ease of construction: 8; After Sales service: 10. Fidelity: 9.

Sources:

Q1 Kit: *South Eastern Finecast, Glenn House, Hartfield Road, Forest Row, Sussex, RH18 5DZ.*
Markits Wheels: *PO Box 40, Watford, Herts, WD2 5TN.*
Branchlines Gearbox + Mashima Motor: *Branchlines, PO Box 31, Exeter, Devon, EX4 6NY.*

Charlie complete: my EM model of 33026 as running in the late 1950's may not be 100% accurate, but to me it does capture the character of these unique engines.

RailMODEL *Digest* Subscriptions

Subscriptions are available worldwide and run for 4 issues. The cost within the UK is £24, or £28 within the EC (including the Republic of Ireland); elsewhere in Europe, the rate is £30. USA/Australia/Japan airmail printed rate subscription is £39.72.

The Digest is now despatched in Mail-Lite padded envelopes to all destinations UK and overseas. Surface rates are available to Airmail Zone 2 destinations. Please note the Digest is now available retail in Australia from Proto-HO Pty. Ltd., PO Box 1534, Orange 2800, Australia and in the USA from International Hobbies, 10556 Combie Road, Suite 6327, Auburn, CA 95602, USA.

Photocopy Subscription Order Form

Please enter me as a subscriber to RailMODEL Digest starting from issue No: [   ]

Name.. Title ....

Address: ....

....

Postcode/Zip: ....   Country ....

Subs. Amount: £...............................   Airmail? [   ]   Cheque enclosed: [   ]

Please debit my credit card. Card type [   ]   No: [   ] [   ] [   ] [   ]

Expiry date: .... / ....   Signature: ....

Tin Tabs, Ebenezers and the Piggery Mission
Tim Shackleton gets religion - in a small way

The Wills chapel is absolutely typical of the genre, in corrugated iron (although wood, brick and stone were equally common prototype materials) and just crying out to be adapted into something else - of which more anon.

Your editor (Rice) and I, it would appear, have more in common than advancing years and an aura of faded gentility. We each live in houses carrying a 1690 date stamp; we both fight a losing battle with the piano (Rice with his beloved Bach, myself in the frenetic barrelhouse style of a New Orleans bordello, c. 1930); and we are interested in the architecture of non-conformist chapels.

Chapels, like locomotives, come in great variety. There are the very grand and important, and there are the humdrum and distinctly shabby. My sympathies are very much with the latter, as are Rice's. Most of your man's layouts – Butley Mills, Leintwardine and Woolverstone spring readily to mind – have featured some humble chapel in plastic tin or cardboard brick. Until he put me right on the matter, I had the uneasy impression that he was some kind of religious zealot, just waiting for the opportunity to thrust a cyclostyled tract into my hand and save my corrupt soul. My own working relationship with t'chapel, I feel I should say, is irredeemably scarred by the crushing tedium of childhood Sunday mornings at Bethesda but as pieces of architecture, I find such buildings fascinating. Where once I sought out lesser-known sub-sheds and signing-on points, now I look for tin tabernacles lurking in the middle of nowhere.

Department of social history
A word of background might put things in context. The nineteenth century was the great heyday of chapel-building in Britain and, though it shouldn't concern us unduly here, the state of Christianity in Britain in this period was an immensely complicated one, a wicked tangle of economics, politics, the class system, social change and many other factors besides the comparatively simple matter of prevailing religious orthodoxies. Outside the framework of conformist religion – which, in the nineteenth century, meant the Church of England and to a much lesser degree, the Roman faith – lay a whole culture of popular religion aimed at the masses.

Within this stratum were the large and comparatively well-supported nonconformist bodies such as the Wesleyans and Methodists as well as all manner of millenarian and adventist sects whose congregations are, at this remove, almost impossible to measure - it could have been thousands, it could have been a few dozen. Many had affinities with particular regions – the Bible Christians and the Plymouth Brethren in Devon and Cornwall, for instance, the Primitive Methodists in Derbyshire, Staffordshire and Salop, or the Particular Baptists in Northants, Bedfordshire and Huntingdonshire.

In the heart of ARC and Yeoman country, traffic thunders along the A361 at Leighton, Somerset, past this rather fine - though long disused - chapel built from the local Mendip stone. In its day it would have served the local agricultural workers as well as quarrymen and their families. Now it stands in a farmyard.

These denominations were, almost without exception, bereft of the immense land- and property-holding resources of the established church. This economic fact is reflected in the frugality and lack of pretence evident in their accommodation. Plain and simple, almost to the point of austerity, their chapels were, on the whole, frequented by people whose lives were not touched by the Church of England. The worshippers in these modest Bethels and Zions consisted largely of the 'labouring classes' and their families, with perhaps a leavening of what we would now classify as lower-middle class people. Often chapels would be very localised in influence, their congregations being drawn from communities that, though still predominantly rural in character, had strong ties to a particular industry, such as quarrying or mining, or where traditional methods of manufacture had not yet given way to the factory system. Many industries – iron working, for one – had developed in areas away from the main centres of population and without strong links with existing churches. Since the High Church all but ignored the lower orders, such fledgling communities were highly attractive to the crusading evangelists of the newer faiths. Some of these branches of the faith were home-grown, like the Plymouth Brethren or the Southcottians, who were disciples of the self-proclaimed prophetess Joanna Southcott – both of these sects survive to this day. Other evangelists came from abroad – the Utah-based Mormons had, by the 1840s, gained a considerable toehold in the Potteries and in industrial Lancashire.

And then there were the church missions, another branch entirely, growing from different roots. Godlessness in one form or another – and the means of counteracting it – was one of the pet hobbyhorses of the Victorians. It was widely believed by the pious that many of the 'lower classes' were deterred from attending church or chapel because they could not rise to the smart 'Sunday best' clothing that was expected. The Church of England, for one, felt that they might be attracted to worship in less ostentatious surroundings. Iron mission churches that were 'tasteful in design, economical and durable' and could be 'taken down, removed and re-erected at a small cost' are advertised in the 1889 Yearbook of the Church of England. There were even crusades to meet the spiritual needs of railway construction workers – the legendary 'navvies' – who were ministered to by charitable foundations such as the Navvy Mission Society and the Christian Excavators' Union. These bold crusaders provided rudimentary places of worship in the semi-permanent navvy camps that grew up beside major civil engineering works such as tunnels and viaducts.

Articles of faith

The tastes of these religious groups, both spiritual and architectural, were a good deal more austere than the elaboration demanded by high church ritual. The earliest of their chapels – those built in the eighteenth and early nineteenth centuries – were often converted from suitable vernacular buildings, such as barns or houses. While some of the later chapels of the Methodists, Baptists and Wesleyans were huge neo-classical or baroque temples of the kind that are so characteristic of West Yorkshire, South Wales and other dissenting strongholds, the lesser sects stuck pretty firmly to the utilitarian tradition. The abiding characteristic of their Ebenezer Chapels and Rechabite Halls was a minimalistic severity of outline and decoration. Most were little more than huts, often of timber-framed construction and clad in wood or corrugated iron, though stone and brick-built structures are far from uncommon. They were well within the capability of local builders working to a price although 'kits of parts' were available at reasonable cost from major manufacturers such as the Butterley Iron Company. The area of land they occupy is usually very tight too – clearly, these were shoestring operations, where every unnecessary expense was spared, and few had a churchyard as such. A considerable number survive around the country, some still fulfilling their original function. Devon and Somerset are particularly good hunting grounds, as is East Anglia. Because of their simplicity and compactness – allied to the fact that it is quite an easy matter to dismantle and resite an old timber or corrugated iron building – many have found further use as village halls, scout huts, sports pavilions and a myriad other purposes.

You will find very little about these modest places of worship in the published wisdom on church architecture. Being keen on these things, however, Rice is threatening to produce a slim but delightful volume of assorted nonconformist chapels under the Hawkshill imprint. Until then, the best source of information that I know of is the *Inventory of Noncomformist Chapels and Meeting-houses in Central England* (Her Majesty's Stationery Office, 1986). This is a hefty tome but, rather like the old Ian Allan Combined Volume, it has been broken down into volumes dealing with individual counties and copies of these turn up not infrequently - and at very reasonable prices - in the better class of second-hand bookshop.

Man with a mission

Having ploughed through several of these volumes in the course of my own researches, I was very pleased when, not so long ago, Wills introduced a kit of a typical corrugated-iron chapel into their Scenic Series. The plastic parts are well mastered and flash-free; they fit together extremely well. Like the rest of the range, the kit lends itself admirably to adaptation into something else. The Gothic arched windows, for instance, are very ecclesiastical in themselves but can be opened out and replaced by something more conventional to give the makings of a small workshop, an industrial building or whatever else takes the fancy.

I built my kit pretty well as it came apart from reinforcing each corner with 60 thou square Microstrip. The drainpipes were a bit on the heavy side, scaling out at about six inches diameter, and these were replaced by 1mm brass wire with the fixing brackets built up out of 5amp fuse wire. The detail of the ironwork was superb, down to the boltheads, but I felt the finials were a bit heavy and so I made my own out of brass rod, chucked up in my mini-drill and turned to shape with high-quality half-round needle files, using a triangular file as a parting-off tool. I made several finials and chose the best. Like many latheless modellers, I feel I'm missing out on something but it is surprising just what can be done in this improvised fashion.

This timber-built chapel near Kersey in Suffolk is slowly falling into disrepair but is still periodically used for worship. Built by a prominent local farming family in the last century, there isn't a churchyard as such, just a grassed-over area that sees the mover (or scythe) only infrequently. The porch is a highly characteristic feature of these structures, whatever the building material. The cream paintwork, with window frames, doors and other features picked out in insipid mid-green, is also very typical, although I recently came across a Methodist chapel in the suburbs of Colchester whose fabric of corrugated ironwork was rendered in startling Deltic blue.

The stone-built Wesleyan chapel at Oakhill is one of my favourite buildings in the whole of Somerset, a lovely design whose domestic proportions demonstrate that simplicity need not always mean austerity. The meeting houses of the Society of Friends often exhibit a similar quiet elegance. There is a subtle ownership code embodied in chapel nomenclature, though space constraints preclude an explanation here - this is, after all, a model railway paper! However, the common name 'Ebenezer' signifies a chapel belonging to the Plymouth Brethren, one of the larger strict non-conformist sects. With paintwork freshly renovated, the Brethren's chapel at Wedmore in Somerset is in absolutely spanking trim, as befits a village that nowadays boasts its own opera. When the chapel was built in the latter years of the nineteenth century, it was nowhere near so grand a place.

By contrast, bleak Pennine moorland is an appropriate setting for the equally grim-looking National Spiritualist Church and Lyceum, high up on a rainswept hillside above Sowerby Bridge, home of the last Lancashire & Yorkshire 'A' class 0-6-0s. It doesn't look a very happy-clappy kind of place, does it? Like South Wales, the old West Riding of Yorkshire was a real hotbed of non-conformist religion in the last century and chapel allegiances run strong to this day.

I airbrushed the completed model with Railmatch Early DMU Green (ref 307), the mid- and dark greens being very common on chapel structures. I wanted to suggest an atmosphere of neglect and so the weathering and general distressing was very heavy, executed using Carr's powders liberally brushed on and off. I copied the patterns of corrosion from the old bus garage in our village, which was also green once upon a time but is now all but 100% rusted. The ridges, which get the brunt of general wear and tear, seemed much lighter in colour than the furrows. These worn-paint highlights can be emphasised by careful drybrushing, while darker paint or powders can be run into the corrugations and left to their own devices. A strong solvent such as Liquid Poly or even a proprietary brush-cleaner can induce a very satisfying crazing of the top coat of paint - this effect is particularly noticeable when, as often happens on roofs, the corrugated iron has been tarred over - but the application needs to be sparing in the extreme to avoid melting the plastic. People can get very sniffy about corrugated iron as a building material but I think it integrates very well once the surface starts to break down and a patina of age has formed. Gently rusting greens and blacks seem to take on a curiously natural quality which I feel harmonises very well with thatch and tile and rendered walls. It is the more 'modern' and 'environmentally sensitive' materials such as stained wood and synthetic brick that, to my eye, stick out like sore thumbs.

Real chapels seem often to fit into awkward little corners – another reason why they are appropriate to a railway model – and mine sits on a triangular site which gives a much more dynamic presentation than the more conventional rectangular plot on which model buildings (unlike real ones) are usually plonked. Chapel grounds are usually plain and simple and so here I have been economical with the landscaping, using Woodlands Scenics products and found materials such as carpet underlay. The fencing and gates are from Scale Link, set on to a half-wall of Howard Scenics brick.

The basic Wills chapel, improved with some better (wire) rainwater downpipes, a pair of mini-drill turned finials and a rather more characterful chimneypot.

The Wills chapel, like the rest of the range, is capable of infinite permutation – I had a Temperance tea room on the go, as well as a narrow-gauge engine shed (two kits butted together end-to-end, with the windows replaced by something more secular) but missed Rice's deadline because I got carried away with the scenic work on the chapel. In creating these dioramas I have a very clear idea of what I want - in this case, the feeling of a rather stark respectability fighting a losing battle with material deterioration and decay (very allegorical!).

I enjoy making these little cameos, whether or not they have any ultimate purpose in relation to any model railway I am likely to build. As I worked through it, this article began to shape up in my head. Rice had been talking about pieces with a broader appeal, that weren't exclusively railway in outlook and attitude but where the modelling element was still strong. I rang the fellow to see if he'd be interested.

'Oh yes,' he said. If he likes what he hears, you get an answer immediately. It's the long silences you have to worry about.

'Should we have a little background history?'

'All right, as long as it's not too heavy.'

'Some photographs of real chapels?'

'Of course. Make sure you get the one at Little Yeldham.'

'I already have. I thought the kit was very good, by the way. Have you seen one yet?'

There was the briefest of pauses.

'I designed it,' said the Hon Member for Chagford, with just the faintest trace of smug satisfaction in his voice.

I sited my chapel on an odd-shaped pocket of land, as they so often were. Often, the land was a gift from a member of the sect - which is why so many rural chapels are in the corners of fields. Farmers were the backbone of rural nonconformism and a great number of chapels stand on farmland or even within farmyards.

*All photos:
Tim Shackleton
(except as noted)*

A classic rural tin tabernacle on a field-corner site at the junction of two minor roads - as was often the case. This Devon example - Taw Green Baptist Chapel - is a long-standing favourite of Rice, who has been photographing them for many a year now. Regrettably, this chapel has now been 'restored' for secular use, totally destroying its unique atmosphere.

Photo: Iain Rice.

Red oxide primer is a good 'chapel' colour, seen here on St Chad's Mission Church at Blists Hill, Shropshire. Built to serve a mining community, this is a true 'tin tabernacle' and the building has now been preserved. The Ebenezer Gospel Hall at Wedmore in Somerset is as fine a specimen of the functional tradition as you will find in chapel architecture. Note the important details which give it its character, such as the windows - very ecclesiastical in outline and similar in style to the Wills model (Not surprising; the kit was partly inspired by this very chapel! - IAR) - and the cast-iron railings, easily replicated in miniature using Scale Link etchings.

A late flowering of the gospel tradition, in this case at Little Yeldham in Essex. Despite appearances, this mission hall dates from the immediate post-war era and the building of Land Settlement houses in East Anglia. Many had a piggery attached – a fine rural tradition – but in this case the pigs left and believers took their place, which explains why the mission stands in someone's garden.

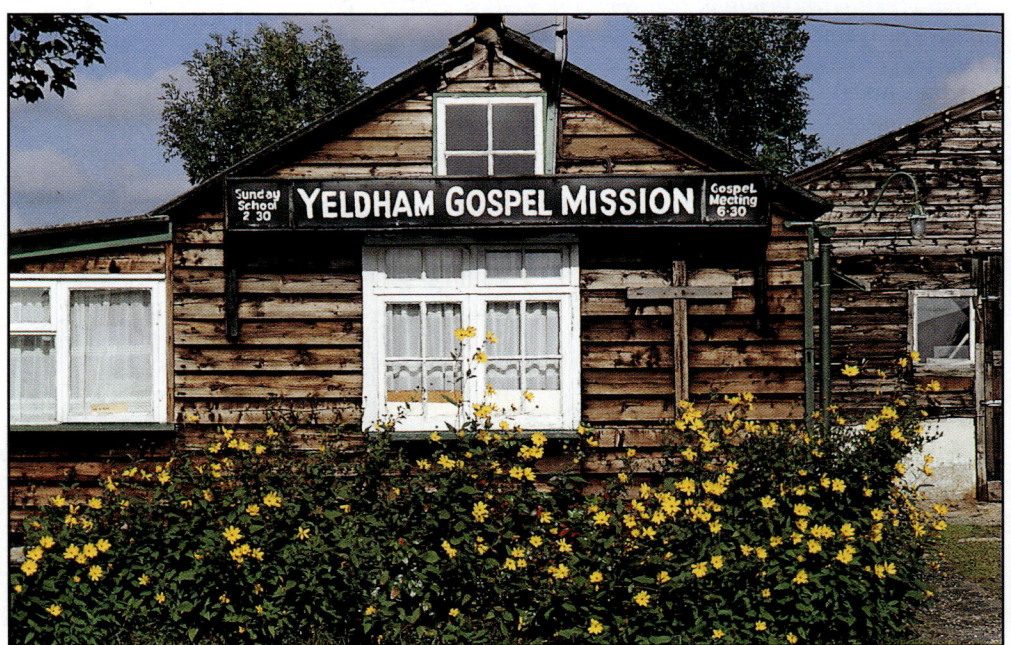

Issue No. 4

ARCADIA

By Martin Brent out of Colonel Stephens

7mm. FS

All Photos: Iain Rice

Not to be confused with that part of Greece that, by strange co-incidence, goes under the same name, **Martin Brent**'s Arcadia is situated firmly in that corner of Kent so aptly called 'The Garden of England'. This is a landscape lovingly immortalised by H. E. Bates as the setting for his 'Darling Buds' - a verdant countryside, rich, lush and fruitful, where the painted strawberries vie with the hops and the apples. It's a landscape that seems perpetually to bask under a balmy summer sky, the verges bright with wild flowers and the hedges bursting with birdlife. All these characteristics are caught to perfection in this affectionate 7mm. scale portrait of a land and a railway that time almost forgot...

No tip for this hapless porter, one suspects; his passenger does not look the sort to see the funny side of her smalls descending into the dust! The figures are by Phoenix - quite excellent.

The 'Tilling Tram' runds into Arcadia after its brief, uneventful jog across the wide open spaces of the Isle of Oxney (the western end of Romney Marsh) from Tilling, that fine old market town on the borders of Kent and Sussex.

*The 'main line' (compared to the Isle of Oxney & Tilling Tramway, almost **anything** counts as 'main line'!) working waits to leave for Codiham (Codgum to the locals) and Appledore. The ex-SECR 'H' class 0-4-4T and stock from the same company have been on this run for the better part of half a century, under three different regimes. Plus ca change... Arcadia has a fine and imposing bracket starter at this eastern end, a signal whose importance seems almost at odds with that of the station itself.*

Two level crossings in quick succession span the by-road from Wittersham and Iden north to Codiham, with Arcadia's corrugated-iron station building between the wharf line of the Tramway and the BR line. The MG VA (1937 model) waiting at the far crossing is, of course, a model of Martin's own VA, entirely fitting. Could it be that errant schoolboy Potts Minor is trying to hide something from Arcadia's very own guardian of the law?

Time for reflection (more haste, less speed..) as the Tilling Tramway's engineering staff tackle (or think about tackling) some minor problem, although whether it is the PW trolley or the Tramway's Hunslet 15-inch tank 'Smug Oak' that is the subject of contemplation is by no means clear. 7mm. scale offers endless potential for beautifully observed and executed detail modelling of this sort, a potential fully realised on 'Arcadia' with a number of similarly delightful cameos.

The Kentish equivalent of Dylan Thomas' 'No-good boyo' comes a-calling on a young lady who certainly looks earthy enough to find herself between the pages of an H. E. Bates short story - though what she's doing somewhere as prosaic as No. 1 Railway Cottages, Arcadia, might take some explaining. The figures are by OMEN, who include some considerably less decorous young ladies in their range.

The 'H" class at the head of the branch train rocks its way across Arcadia's pointwork as it rolls into the platform with an Apledore-bound working. The painter up the ladder on the signal-box was a primary spur to Martin's entry into 7mm. scale - truly realistic figures are a huge bonus of the larger size.

*And Arcadia's just another station
on his twice-a-daily
pere-
grination!*

Preamble: Arcadian Origins

The Brent household is not noted for its belief that progress is necessarily a good thing. We are just getting used to the lack of cardboard milk bottle tops and regularly bemoan the lack of an electric gramophone on which to play our Al Bowlley 78 r.p.m. records. Similarly, light reading has not been the same since the untimely demise of two leading proponents of the light novel, Mr. E. F. Benson and Mr. John Moore. As for motoring, we are firmly of the belief that any worthwhile development of the motor car ceased around September 1937 at Abingdon-on-Thames. In short - we do not like change. Hence the reason for foisting yet another layout rejoicing in the name of Arcadia on an unsuspecting public.

Those gentlefolk who used to read the erstwhile and lamented magazine 'Model Railways' and have long memories (or a goodly collection of back numbers) will recall that about the time that our worthy editor was producing his magnum opus on 4mm. driving wheels - March 1981 to be exact - there appeared a modest offering relating the tale of the writer's cautious dip of the lefthand big toe into the, for him new, waters of EM modelling. This small layout was given the name 'Arcadia' - which hamlet really *does* exist, in the wilds of Kent just to the north of Tenterden. Pre-1954, it was served by the Kent & East Sussex Railway - and we all know who built *that* don't we......

The redoubtable Colonel Holman Frederick Stephens' creation must have caught the eye of the editor of 'Punch' because in the edition for 3rd June 1946 there appeared a delightful poem, illustrated by the incomparable Rowland Emmett. The Farmers' Train - a pastiche that truly caught the atmosphere of one of the Colonel's light railways. With verses such as:

'Ever seen a railway train
wheel deep in the wheat?
Poppies on the boiler dome:
wreaths of bitter sweet
twined about the driving wheel-
burnished brass and polished steel:
puffs of steam like woolly lambs,
on the line to Bodiam?
(Punch, 3rd. June 1946)

What else could one do but fall in love with the poem and the railway it so wittily portrayed? And so the name 'Arcadia' was appropriated and the first (tiny) layout with that name was born.

Antecedents and Developments

After a few years - and some 20 or so exhibitions - we began to feel the need of something a bit more interesting, so Arcadia was extended and changed; so much so that it was necessary to rename it: Enter Rye Harbour. Eventually this layout, too, ran out of interest - and was last heard of as a coin in the slot display in Brecon!

In the meantime, there had arisen a much more interesting layout, Hope Mill, again in EM, but this time a double track passing station based loosely on the old Elham Valley line. To avoid boredom, we ran a train service more suited to the Reading-Redhill line than a sleepy Kentish by-way. This layout was a joy to operate - full of interest and with virtually unlimited potential so far as train movements were concerned; trains ran through either as expresses or stopping as required; trains terminated, split, shunted and met one another (in the nicest sense of the word!). The branch trains ran in and out and, with no bay, had to keep diving into the goods yard to escape the nasty, over familiar, attentions of those great big, bullying, mainline trains. All of these train movements made for very interesting operation.

In time this layout too was extended, to include Tilling Junction - where the branch line for Winchelsea Road and Marston Abbotts came in.

By this time it was some 30 feet x 3 feet and required a team of six to operate it to its full potential. Great fun but we could only set up the whole shebang at exhibitions.

Change at Last...
However, even in such a reactionary household as ours, things *do* change and it was eventually decided that the M 25 being built virtually at the end of our valley was a bit too much. We decided to escape from the frenetic activity, the noise and the traffic fumes of Hertfordshire to the clean air and tranquillity of Herefordshire. (Well, it only meant changing one letter in the County name; change is all very well, but - as I said, we do not like too much of it!)

It was evident that Hope Mill would not fit in the new house. Also, in the meantime, I had given up being the Sir Humphrey Appleby of Watford and had taken up loco kit building as a more-or-less full time occupation. As a result of this, I had fallen hopelessly, head over heels in love with 7mm. scale modelling. Such are the foibles and frailties of middle age!

Having thus built a couple of 7mm. K.E.S.R locomotives as commissions, I bowed to the inevitable and decided that Hope Mill would have to go. In its place would arise, phoenix-like, another light railway - but this time with considerably greater potential for operation. Many pleasurable evenings were spent distilling some thirty years of mistakes in the layout planning field until something that I thought feasible was arrived at. It had to be a through station - and a junction to boot. To give interest, the junction would be between a British Railways single track branch line and a Colonel Stephens line.

The rationale behind this idea was that, pre-nationalisation, there had been a previously-unknown twig of the Colonel's empire, the Isle of Oxney & Tilling Tramway (shades of the West Sussex Tramway!) - only *part* of which came into the British Railways fold. It was presumed that - probably because of a bureaucratic bungle - that part of this line from Arcadia to Tilling was not included in the 1948 Transport Act (rather like the Tallylyn Railway) and so the Colonel's successor, Bill Austen, decided to hastily collect a motley equipage of stock from the other lines to run this last outpost of the late Colonel's empire. This, of course, would allow us to build virtually anything from the Colonel's heterogeneous assemblage of locomotives - but it has to be admitted that even with the latitude allowed by the presumption outlined above, we have tampered even more with history to allow the use of locomotives which had been scrapped well before the war or to create locomotives that never existed as part of the empire. No excuses are offered for this laxity; gone are the puritanical rantings regarding prototypical fidelity that were prevalent in the 1960's. Advancing years, a modicum of vintage port and sufficient Stilton have mellowed the outlook more than somewhat.

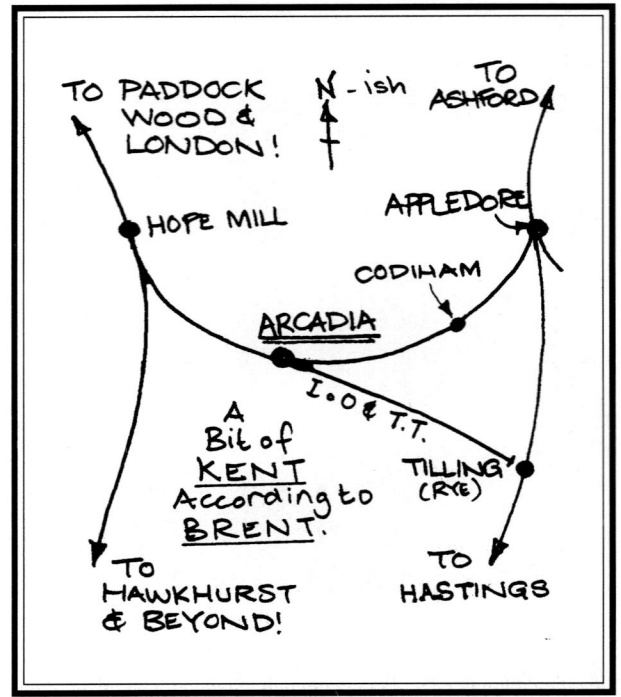

The New Arcadia Arises
The layout as finally devised meant that it was a station where passenger trains could pass, while in the yard wagons could be shuffled to and fro. The light railway trains would come and go and there would be a general interchange of traffic. Sounded good; all we had to do was build it.

This took a lot, lot longer than anticipated, partly because of the move of house and partly because I found that 7mm. scale modelling was so satisfying that I spent a lot more time adding or creating detail than had been possible, desirable or even practical in 4mm. scale. This was most evident in the fields of landscape and architecture, where I found that modelling in the larger scale was a whole new ball game. Of which more anon.

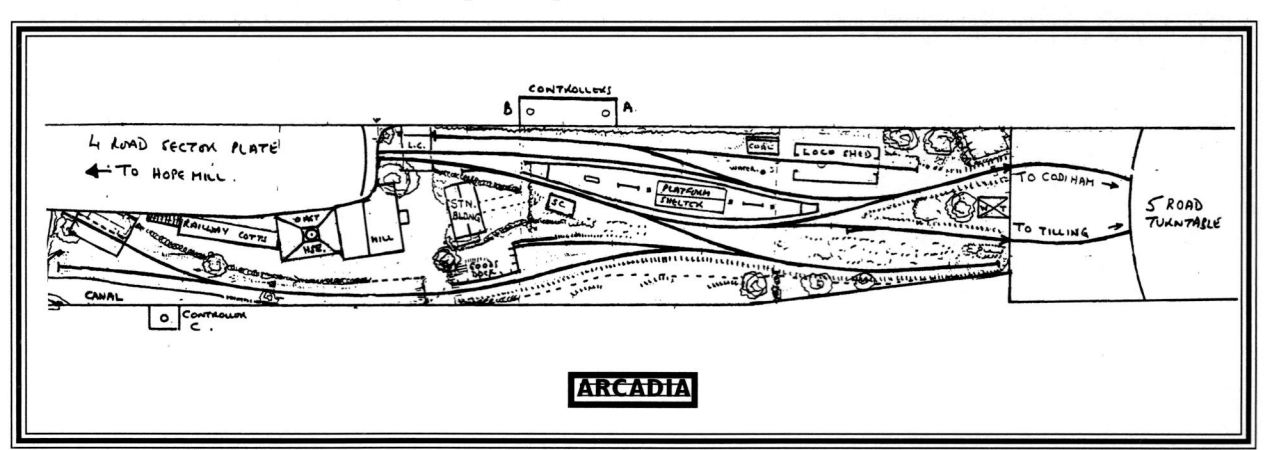

A train runs into Arcadia on the BR lines, passing a wonderfully-observed clutter of platform-end details and, of course, Miss Elizabeth Mapp having a spot of bother with a clumsy porter. A simple scene caught to perfection.

It can be seen from the plan that there is a single track line which is operated by B.R. In our imagination it runs from Hope Mill to Codiham (another fictitious Sussex village immortalised in print by Donald McCormick under the title, 'The Wicked Village'). This line loops around an island platform a la Merstone or Haven Street, both Isle of Wight prototypes - in fact, much of the layout owes a large debt to the former station. At the right hand end the light railway - the Isle of Oxney & Tilling Tramway - wanders off to the small town of Tilling which is, incidentally, served by both the Tilling Tram and an ex-S.E.C.R. line.

To complicate things, the light railway locomotive shed is one side of the B.R. line while the very basic goods yard is t'other. Also, from this yard which is open to both 'companies', there runs a line - still the sole province of the light railway - to a small canal wharf. 'Unlikely in this part of Kent/Sussex', you might say. 'Quite possible' says I - who used to drive home from Rye past the rotting lock gates on the arm of navigation that left the Rother and headed off towards Brede. Anyway, Jan and I love canals and escape to them whenever possible - and it's our layout!

Portability

The layout was devised to be transportable as we do enjoy making an exhibition of ourselves. It was therefore divided up into five separate baseboards. Four are 40ins. long by about 30ins. wide, while the fifth is a bit longer - some 48ins. or so with a turntable type fiddle yard 54ins. in length protruding over the end. This gives a total length of about twenty feet. In case this sounds a bit vague it is probably worthwhile explaining that originally the 48-inch board carried a fiddle yard utilising cassettes but we found (at our first outing to the oh- so-delightful Southwold exhibition, which we try and attend each year) that, due to the height of the layout and the length and weight of the cassettes, this system was more than somewhat difficult to use. It was therefore decided to revert to the tried and tested turntable

Baseboards

As well as having a high resistance to change factor (a management phrase all the rage when I was involved in work study) the Brent household is nothing if not frugal in some respects. The baseboards were, therefore, built from redundant ply packing cases which were left on the tip at the office after the new computerised telephone system was delivered. These cases were built from half inch, waterproof ply which was swiftly converted into baseboards, built and braced using strips of ply arranged in a triangular formation so as to give rigidity and resistance to flex or twist. The side members are 7ins. deep, partly to give rigidity but also to protect the equipment that is slung from the underneath of the boards, notably the 'Tortoise' point motors which are some 4ins. in length.

The backscene, made from $^1/_8$ in. hardboard is braced with softwood strips and screwed and glued to the back of the boards, again to give rigidity. As an aside, the next boards that I was involved in building were those for the Watford Club's 'Smug Oak' 7mm. layout and by that time we had devised a method whereby the backscene was integral with the vertical board members and formed a hollow box section which could be used as a wiring conduit or for point motors or levers. Inter-baseboard alignment was by means of EM Gauge Society pattern-makers dowels and quarter-inch bolts with wingnuts.

All the baseboards for Arcadia were supported on two long strips of 2ins x 1in. timber - hinged for transport - which, when in use, mount vertically in three slotted trestles. The board ends are also slotted so that each board sits on these strips. It will be appreciated that using this method, it is easy to put the layout up roughly and then slide it all together for bolting up. When assembled, Arcadia stands some 40ins. off the floor; a bit high, perhaps - but the view for the average man or woman is improved and, more importantly, it gives a lot more storage space in the workshop that would be lost if it were to be mounted lower.

Trackwork at Arcadia is not exactly up to main line 'prize length' standards, but it is very typical of the sort of PW that would have been found on a BR byway in the 1950's, and it adds much to the veracity and atmosphere of the layout. Components used include PCB sleepering together with plastic chairs and timbers from Peco and C & L, with some Slater's chairs here and there.

There is a snag to this high-level presentation, in that vertically challenged operators (anyone below about 6 feet) are hidden behind the back scene which towers some 12ins. or so over the base boards. 52 inches is a bit much for anyone to reach over - unless they are well over six feet in height. A stout ply box - we like to refer to it as a 'podium' - has provided a workable answer.

Trackwork

The track is a bit of a mixture. The ordinary track is a mixture of Peco and C & L while some of the points are from parts or kits from both these manufacturers. The first experiments with plastic based trackwork were not encouraging. If I had bothered to read the instructions things might have been different but, being insufferably complacent - even smug - reading instructions was, I thought, a bit beneath me. After all, I had got away with this so many times before in 4mm. scale; why bother in 7mm.?

Suffice it to say it all went wrong and I became prejudiced against plastic track in 7mm. scale. Hence, there is a third type of point in use, based on copperclad glass-reinforced PCB sleepers which I knew all about without having to bother about tiresome instructions. To give the effect of the rail raised by the chairs above the sleepers - which is so noticeable on the prototype - a small packing piece was interposed between the rail and the sleeper. This was done by using paste flux to hold it in place during soldering. This may sound cumbersome, but in fact it was quite easy and this form of track construction was quick, with a very strong finished product.

After the point had been built and checked, cosmetic chairs by Slaters, C & L or Peco were superglued to the rail and clearances carefully checked. Any of the inner chairs protruding above flange height were trimmed with a very sharp craft knife. Most of the points were of a nominal radius that would terrify some of my 7mm. friends. 48ins. was common and there was a curved point in the yard that I have not yet dared to measure! When laying the track, a reverse curve was unintentionally introduced on the main line, adjacent to a point, which caused problems with buffer locking; the only satisfactory answer was to lift and relay the affected track. Once again, plastic components were tried, with a reading of the instructions that helped immensely. This time, I got it right. My prejudices disappeared overnight.

It has been remarked by certain sages that the points appear to have been placed anywhere on the layout with no regard to avoiding baseboard joins. Unless one chooses to work to exact scale standards, I am strongly of the belief that there is no need to worry - providing the points are laid across the board joins and cut with a rotary slitting disc after all is secure. That said, every effort was made to ensure that the crossing-isolating gaps coincided with the board joins, so as to avoid two gaps within a couple of inches.

Ballasting

The track was laid on $1/8$ in. cork and ballasted with 4mm. granite, mixed with a brown scatter and a modicum of green scatter to represent grass at siding ends. This was spread dry and, after tidying up the edges, an eye-dropper filled with a dilute PVA-with-a-drop-of-washing-up-liquid mix was used to fix it all in place.

For the overgrown parts of the light railway a different technique was used. The track was laid quickly into a bed of PVA and, before fixing the track, the bed was sprinkled with a drop of spent ballast to give a patchy effect. The whole area was then 'puffed' with Heiki or Noch electrostatically charged flock to give an overgrown, grassy effect. It has to be said that it is necessary to work like greased lightening as the PVA quickly forms a skin and the grass will not adhere to it. The overall effect is, however, worth a little bit of hyper-activity.

For the all-but-engulfed canal wharf, yet another technique was used - carpet tiles of the long hairy pile type. The sleepers were embedded in the grass and, to achieve this effect, the carpet was literally branded, using a soldering iron to melt the top surface and create the slots for the sleepers. To facilitate this, a tool was originally made, consisting of a steel bar brazed to a Brewster 75 soldering iron tip. But it was found that not enough heat was generated, so in the end each recess was individually carved out of the carpet tiles with an ordinary soldering iron. The carpet was sprayed with an airbrush loaded with a yellow/green mix before track laying. Different areas had different numbers of coats to give the desired effect. Evostik was used to locate the track and the pile was then raised with a stiff wire brush. The result was more than somewhat pleasing and has been the cause of much comment. It was also very quiet.

Operating Considerations
As already mentioned, operation is important and after initially running the layout with two operators, it became evident that it would be advantageous to have a third person at the front of the layout shunting the yard virtually full time so as to make things more interesting at exhibitions. This would also get over the problem of the height of the backscene impeding rear operation.

Adding this facility involved making fairly major changes to the basic wiring already installed and was the subject of much thought and scribbling on scrappy bits of paper. While the front man is shunting to his heart's content, the two main operators drive the trains on the B.R. line and the light railway and operate the fiddle yards nearest them. Generally, they only drive trains towards the yards for which they are responsible. That way they can see what they are doing. To do it the other way, to ask someone to drive a train away from them and to stop it with precision where it can't be seen is a recipe for disaster. The effect of asking operators to operate fiddle yards as well as drive trains is a bit like swans-a-swimming - all serene on the surface but frantic activity where it can't be seen! While the view that the spectating public have of the operators above the backscene may look relaxed, there is a lot going on out of sight!

Highly effective grass-grown track on the wharf branch was achieved by 'inlaying' the PCB-based FB track into hairy carpet tiles, which were painted with an airbrush. Note the pheasants and other fauna lurking at the base of the hedge, and the tall clump of wild flowers - cuckooflower? - growing amid the grasses.

In fact, we have found that this is a far more demanding layout to operate than Hope Mill, mainly because of the single track, short loops, short trains and even shorter headshunts. It is a layout that is very easy to snarl up - but it is also great fun, which is the main thing. A series of train schedules as distinct from a timetable are used to make the most of the operating potential.

With the low speeds and short trains, the AMR handheld controllers (intended for 4mm. scale) from another layout have been pressed into service. These are fed from individual 18 volt transformers which are separately boxed and mounted on the floor, as described in my first layout wiring article in *Digest 1*. Separating the power supply keeps the 240 volt mains away from the layout and, since there are some seven transformers providing the urge for traction, points, signals, uncouplers and accessories, the weight of this box is quite considerable. It seemed only prudent to mount it on the deck. It also made good use of a rather nice box that was purchased from Proops for a fiver!

The obligatory Terrier (Arcadia has a breeding pair) runs in from Appledore on the branch passenger train. Small locos like this with short, light trains enable small handheld controllers like the AMR to be used without problem, in spite of their limited power-handling capabilities.

The somewhat dubious timber and feed businesss of messrs. Bossom and Overbury is carried on from this tin shack, lurking at the end of the wharf.

Controls

The control panel takes the form of an old telephone operators' console (Proops, for a fiver, yet again!) inverted and bolted to the layout. My son-in-law, who is an industrial engraver, produced a lovely engraved plastic signal box style panel complete with all lettering and the switch holes pre-cut. The panel is designed for use by two operators who can select their respective routes, uncouplers, etc., but if we have a spare operator, he or she is often dragooned into acting as signalman. Cab control is used and the five sections can be switched to either of the main controllers while the goods yard can be switched to the third controller when necessary.

Point control was effected by single pole switches operating Tortoise motors. Quiet and powerful, initial impressions are that these motors are the 'bees' knees' - although to cope with the inherent stiffness of 7mm. rail it was found necessary to beef up the operating wire. Another time I would probably pivot the switch rails.

Signals

Signals utilise the evergreen Post Office relays with gainstroke arms fitted. All signals work, including the S.R.-pattern Westinghouse ground signals which were home brewed. In fact, the bulk of the signals were scratchbuilt, using Model Signal Engineering arms, finials, caps, lights and cranks and Scale Signal Supply ladders. The rest of the signals were made from oddments: Posts are of stripwood - old rocket sticks collected after 5th November - that frugal, nay some would say tight, streak surfacing again!, while thin sections of large diameter tube parted off in the lathe and 4mm. scale rail were pressed into duty for brackets on the three-doll starter. Handrails are wire, with scrap etch strip for stays and thread for guys.

Landscaping

The landscape was the next task and, to be honest, this was the bit we enjoyed most. Very careful thought was given to this - there was not a lot of room and it was important not to make the layout look overcrowded. It was also necessary to hide the exits to the fiddle yards. The idea of tunnels or bridges did not appeal and so, at one end, the trains run over a level crossing and behind Hope Mill - a large Victorian edifice owned by one Jno. Cox, a local purveyor of corn. To assist the illusion that the trains are simply disappearing behind the building, a couple of large trees and a hedge are used to break up the scene from most angles.

At the other end, the main line runs off-scene to Codiham past a couple of smaller trees and a typical Col. Stephens water tower, a simple tank set on a sleeper base. Nearer the viewer are a couple more trees which again add to the effect - they also serve to hide the fact that the light railway is quietly sneaking offstage, again behind trees and hedges. In the foreground of each end of the layout there are a number of small cameos - an animated fisherman who occasionally wakes up enough to see what is on the end of his line or again a group of rabbits at play. Anything to distract the eye from the 'wings'.

Largest and most imposing signal on Arcadia is this 3-doll bracket starter, concocted from spent Guy Fawkes rocket sticks, MSE arms, lamps and finials, Scale Signal Supply Ladders and oddments of brass, tube and so on.

Landscape Construction

The landform base was carved out of 50mm. block polystyrene left over from insulating the railway room (an extension to the garage), and this was covered with old towels which were soaked in dilute PVA. When dry, the roads and loading banks were created using plaster tinted with water colours, smoothed off with a wet household paint brush. The rest of the vegetated areas were given a basic covering of fairly coarse scatter materials.

Trees, made in the main from C & L kits, multi-strand wire and tape with foliage from rubberised horsehair and Woodland Scenics foliage matting, were planted into the styrene land-base. The buildings were embedded in the landscape by the simple expedient of cutting their 'footprint' out of the by-now-stiff towelling. The glue used to fix the buildings was allowed to squidge out around the joins, when it was quickly given a puff of electrostatic grass from the puffer bottle.

The undergrowth, individual plants (by Scalelink) and long grass - from a variety of sources - was now added, and my good friend John Cox and I spent a very happy day adding these various types of plant growth to the layout. In the meantime, Jan was sitting by the pond painting figures, sheep and a variety of animals for all she was worth, and these were fixed in place as soon as they were dry. There is still much to do in this area - there are a couple of $1/43$ rd. car kits waiting to be built and some work has been done towards a number of animated figures.

Backscene

The backscene was dealt with after the basic scenery was in place. Over the years I have heard a saying: 'Never let an artist near a backscene - they think of it as a picture'. Well, the backscene on Arcadia is definitely no picture! It is deliberately vague, made up of a series of blodges which are intended to confuse the eye by removing any sharp lines where the landscape meets the backscene. Any resemblance to trees and fields is not entirely accidental, but the point must be made that the backscene is an integral part of the layout and should seek to complement it - not overwhelm it. *(It's also important to ensure that colour values used on the modelled layout are carried through into the backdrop, something Martin has got 'just right' here - IAR).*

Buildings

Some of the architecture has already been referred to, but a brief explanation of the origins of some of the buildings and their construction might not be amiss. The big brick mill, as already described, serves to hide the entrance to the 'Hope Mill' end fiddle yard. This model was based on a photograph of a mill building in the Medway valley - but at Arcadia it has attached to it an 'Oast', a square kiln used for drying hops. This type is more common in Kent than in Sussex, where oasts tend to be of the Roundel type. My model is actually based on a Herefordshire Hop Kiln (note the different terminology), and the run of buildings continues with a couple of cottages based on some seen on the Ellesmere Canal during one of our holiday narrowboat wanderings.

All of these structures are built from foam-core board suitably braced and covered in the Howard Scenics card brick. This was glued to the building and, when the windows had been cut out, coloured with wax crayons after the entire building had been given a wash of pale grey water colour to represent the cement courses. Individual bricks were picked out and, when satisfied with the result, the model was sprayed with a matt fixative/varnish. The windows are American Grandt Line products - beautiful examples of what can be done with plastic moulding.

Near the mill is a covered loading dock for those importers of dodgy timber, Messrs Bossom and Overberry. This is a 4mm. Ratio Carriage Shed, ideal for a small covered dock in the senior scale. (Apropos nothing at all. (If you're wondering where all these unlikely-sounding business titles come from, it should be noted that great liberties have been taken with the names of the regular team of operators...)

A view across Arcadia from the loading bank gives an idea of the subtle variation in landscape contours, as well as the effective way in which the (very simple) backdrop blends in with the modelled foreground.

The platform shelter at Arcadia is a model of that from Canterbury Road, East Kent Railway. It is built in Plastikard using board-by-board construction to overlay a basic sheet shell. The cast-concrete platform edging is cast in resin from home-produced moulds.

The station office is half of the building that used to stand at Wittersham Road on the K & E.S.R. That too, was at right angles to the track. It is made from Wills 4mm. corrugated asbestos sheeting - which is ideal for 7mm. corrugated iron. Nice clean lines, and even the nailheadss are shown - lovely stuff. Windows are again by Grandt Line. The platform shelter is based on that which stood at Canterbury Road on the East Kent Railways. This is scratchbuilt of plasticard, using individually-cut planks over a 40thou. sheet plastic shell, with the brick base from Slater'sPlastikard brickwork.

The locomotive shed, on the other hand, bears no resemblance to anything on the Colonel's railways; it is a Heljan kit, available from Kittle Hobby, slightly doctored and weathered. Despite its origins I am more than pleased with the finished article. The water tower has already been mentioned. It is a Duncan Models Tank on a pile of balsa strip wood suitable stained with the contents of the brush cleaning jar.

The platform facing and that of the loading dock is based on the standard Southern railway concrete edging. As none was available commercially when the layout was under construction, a plasticard master was made and a rubber mould created from that. The actual moulding is in casting resin - which can be curved while it is curing.

In 7mm. scale, I've found it necessary to give greater attention to elements like down pipes, gutters and building detail. The range of bits and pieces produced by such as S & D and Langley Models are indispensable. The various styles of chimney are brass turnings made from drawings of chimneys found in a local architectural salvage yard.

All buildings are toned down with a thin coat of weathering from the airbrush and the same tint is allowed to wander down onto the adjoining landscape, again so as to avoid a sharp line.

Scenic Details

The fully-rigged wire fencing, electricity and telegraph lines have attracted attention. The fencing posts are standard Slaters products, but the wire is a fine Lycra thread as sold by Fourtrack Models. This is, I believe, used mainly by aero-modellers for rigging the bracing wires on their large scale bi-planes. The same thread is used for the overhead electricity poles and the telegraph wires. Where they cross a baseboard joint they are simply looped around the insulators on the adjoining boards. This lycra is extremely elastic, and even if caught by a straying sleeve these fine wires will not be damaged.

The detailing of the layout gives much pleasure and the range of Phoenix people and accessories have been used, hopefully to good effect. The fauna in this part of the country is amazing: Squirrels both red and grey, rabbits, pigeons, buzzards, pheasants and blackbirds abound. Some are by Duncan Models, some by Scalelink while the smaller rodents and birds are of my own manufacture. Much work remains to be done in this field.

*In **Digest** 5, Martin will be describing Arcadia's unique collection of locomotives and rollings stock - of which the Colonel would have been proud!*

New RailMODEL book coming soon from Martin Brent:

A Handbook of
7mm. FINE SCALE
Railway Modelling

96pp. A4, 8pp. colour, profusely illustrated, a practical guide covering all aspects of working in the senior scale. Contains GOG fine scale and ScaleSeven standards in detail.

Price: £9.95. Available at GOG Spring Fair, March 1997.

Modelling in CARD
- the Stuff of Subtle Structures?

Leintwardine Station Building, a not-very-accurate model of Kinnerley on the Shropshire and Montgomery Railway. Brickwork is embossed into the card - $1/16$-in. thick Apsley Board - and the roof has paper slates over postcard. Painting is in acrylics. The platform, too, is embossed card, overlaid on a structure of thick backing board.

All Scales

As a preamble to a series of articles on structure modelling in card and other paper-based materials, editor **Iain Rice** takes a look at the basics of card modelling, exploring the properties of the medium and considering the few tools needed to work it.

Introduction

Card is one of the most versatile and traditional of materials available to the modelmaker. In recent years, it has tended to be overshadowed by more modern, less organic alternatives, most notably plastic sheet. Yet card, available in many different forms, still has a great deal to offer the modelmaker, particularly in the field of architectural modelmaking. It has unique properties that the newer materials can't equal, particularly when it comes to the representation of those subtle waywardnesses so typical of older vernacular buildings.

Consider the manifold advantages that card has a structure-modelling material. It is cheap, easily-worked, extremely versatile, and surprisingly robust. It takes paint well, can readily be joined to a wide assortment of other, complementary, materials. It can be bent, folded, embossed, distressed or stiffened with the likes of Shellac or PVA. Laminated, it can produce structure shells that combine subtle surface detail with sturdiness. Wet it, and you can mould it to shapes that few other materials can cope with. It is, if finished properly, stable and unaffected by ultraviolet light, that great underminer of plastic-sheet models. I can be obtained in a huge range of thicknesses and grades, often for virtually no cost. Look around, and you'll find a card that will answer virtually any structure-modelling requirement.

Of course, it's no use pretending that card is some sort of modelling panacea, without drawbacks or limitations. Obviously, there are respects in which it can't compete with newer alternatives like plastic sheeting. You can't engrave card like you can sheet styrene, and you don't have the instant-bonding facility that solvent cements achieve. (Although neither do you have the potential for disaster when that solvent goes astray!). And, of course, card - being an organic, 'breathing' material, needs protection against moisture if it is to remain stable and keep its structural integrity. Like any organic - and most inorganic - materials, in other words, it does need some understanding of its properties if you want to get the best out of it.

Some of these limitations can be overcome or mitigated by combining card with other materials such as paper, sheet and strip wood, plaster, modelling clay and, of course, metal and plastics. One on the most useful card-compatible materials to arrive in recent years has been foam-core board, sold under the trade name of 'Kappler Board' and available in a range of thicknesses from 3mm. up to about 15mm. This consists of two thin card skins enclosing a core of closed-cell polyurethane foam, making a material that is light, rigid, robust and easy to work - and highly compatible with card.

A number of specialised card-based structural-modelling products are also on the market - although some of these, as is so often the case, are merely a reprise of what has gone before; compare Howard Scenics splendid 4mm. embossed-card brickwork with that produced - more years ago than I care to admit to - by Ballard Bros. There are some genuinely new innovations, like Exactoscale's 4mm. scale tex-

Hand-embossed brickwork on Apsley Showcard was the method used for this model of a Gloucestershire wayside chapel (I've forgotten where it was!) which I included on my 4mm. light-railway fantasy, 'Leintwardine'.

This texturing was applied with a blunt scriber - the tip was rounded on a slipstone - for the horizontal courses, while the 'perps' - the vertical joins between bricks - were put in with a small jeweller's screwdriver. A little tedious - but very rewarding. The result was painted with a mixture of Humbrol hobby acrylics and designer's gouache.

tured screen-printed brick sheet. Other old favourites like brickpaper have been newly-refined; the Howard Scenics papers are full of a subtlety and a beautifully-observed veracity that was often lacking in the traditional designs from the likes of Merco (although these undoubtedly had their charms). Skilfully used, with a bit of weathering or other monkeying-around to kill any ink-sheen, and by imparting a certain amount of relief detail should the subject warrant it, very realistic results can be obtained.

Surface Textures

Realistic surface textures have long been one of the chief advantages of card as an architectural-modelling medium, and these textures have traditionally been achieved in a number of ways. The first is embossing or incising - the imparting of relief detail to the card by impressing with a suitable implement. This calls for a card, such as the Apsley Board range much favoured by Pendon (Apsley Pasteboard) and yours truly (Apsley Showcard), which has a soft centre but an outer skin that deforms without tearing. These types of card also take paint well, although in a very different fashion, as we shall see.

The second technique is overlaying, possibly the most striking example of which has been chad brickwork, individually-applied paper or thin card bricks that give a delightfully-natural texture and irregularity to vernacular brickwork; 'engineering' brickwork, as found on many railway structures, was rather more regular and of less pronounced texture - for which the embossed-card brickwork is ideal. Stone walling and timber construction - especially weatherboarding - can also be effectively reproduced by overlaying, as on my model of Arkinstall's Mill at Butley, illustrated at left.

Overlaid texturing in considerable variety was used to replicate a number of different building materials on Arkinstall's Mill. The brickword uses computer chads (a bit overscale in 4mm.) while the narrow weatherboard is cartridge paper and the heavier board is postcard. Slates are bank paper (they're thinner than you think!) while the tiles are Wills plastic sheet - cheating?

I built his vintage card model of a pair of Birmingham & Fazeley canal lengthemen's cottages back in the early 1970's using incised brickwork - cut into picture-mounting board as a narrow V-groove with a sharp scalpel; odd bricks were 'lifted out' with the tip of the scalpel to expose the slightly 'furry' texture of the card, a good way of representing 'spalled' brick (brick where a crack has allowed moisture in, leading to frost disloging the outer face of the brick). The roof is, once again, Wills sheet, and was added when Bob Barlow reworked the model to a 'Suffolk' version for inclusion on Butley Mills.

The third method is a variation of the second, in that it is basically another form of applied texture, except that this is applied in liquid form with a brush - as in the use of a plaster slurry to represent render or diluted PVA followed by scouring powder for pebbledash - and in paste form, as when modelling clay, cellulose filler, gesso, acrylic paste or plaster are applied to build up textures, add flashings, cement grouts, corbels or similar details. More on all this in an article of this series devoted solely to texture and finishes.

Structural Properties of Card.
Model structures are, like their prototypes, in need of true structural integrity. They need to have enough inherent strength to remain in the form desired, to support themselves and to be reasonably robust to withstand the rigours of model railway operation. Getting a strong and stable card structure depends on observing three basic fundamentals: Using the right card for the job in the first place, suitably sealed to protect it from moisture; adding adequate and appropriate bracing, either using more card, or with other materials like stripwood or fabric scrim; and using the right glue to hold it all together.

There are, basically, three main forms of card - laminated sheet, cored (usually corrugated) board and pressed pulp. As with most laminated materials, the former has the greatest strength, and is the composition used for most high-quality cards, the types of most interest to the modeller. Laminated board can be 'peeled apart' - a very useful technique in modelling some types of surface detail, as we'll see. Of the cored boards, the most interesting is 'Kappler' foam-core, the ideal material for building the shells of buildings - especially large structures - for overlaying with surface texture material. The last type of board, the 'pulp', also has important uses as, unlike the first two, it has no pronounced 'grain' and will accept compound curving without difficulty. The two commonest sources for modellers are cornflake packets and the backing board supplied with subscription copies of MRJ, for those of you who receive that august organ through the post.

Choosing the right sort of card for the right role within a model is half the battle. Generally, unless you're going for a high-quality coated board like the Apsley Showcard, which has both a fine surface that takes embossing and paint well (it's intended for

Dutchman Renier Hendriksen has forsaken the shiny plastic-kit Faller buildings beloved of so many continental modellers and adopted Pendon card-based modelling techniques for his Cornish narrow-gauge railway. This row of buildings in the village of Moor's End makes good use of the freedom offered by embossing to reproduce some very effective coused stonework - stonework that is both unique to the model (not the same as everyone else's) and also entirely fitting to the actual structure.

The mill at Butley Mills - based on Heybridge Mill, near Maldon in Essex, and until I got going on the Trerice clay dries, the largest card structure I'd attempted. This has a shell of picture-framer's mounting board with 'eggbox' bracing and some interior cameo modelling. Structurally, it has proved strong and robust; building it was a very satisfyng - if time-consuming- exercise.

high-quality artwork) and enough thickness and strength for structural purposes, it's worth considering a robust material like, say, backing board (the thick grey card used by picture-framers to stiffen the material of the picture), mounting board or Kappler for the shell of the building, with a surface of either thin card (such as the Apsley Pasteboard so beloved of Pendon, or the commoner 12-sheet Bristol Board) or even paper. On the large clay dry which forms the centrepiece of my slowly-evolving 'Trerice' diorama I have adopted this approach, with a shell of 5mm. Kappler overlaid with a mixture of Apsley Showcard, Howard Scenics brickwork, embossed 'Cotman' watercolour paper, American corrugated foil (Campbell), cartridge paper, postcard and bank paper.

The pulp board is less useful as a structural material, but is invaluable for achieving irregular shapes such as sagging roofs, bulging walls or curved surfaces generally. I also often use it for 'secondary structural' purposes, like internal partitioning or as a sub-base for applied surface materials - again, particularly roofs. My family have to eat a lot of cornflakes when I get the structural or scenic-modelling bug!

Complementary Materials

There are a number of other materials that work particularly well with card in performing certain specific structural functions. The most obvious is stripwood, about quarter-inch-square, which is an ideal means of strengthening corner joints and bracing long runs of wall where no floors, partitions or other internal bracing are desirable, such as on a loco shed or other large 'open' building. Another item not often used - but capable of imparting great strength - is plasterer's scrim (or an old gauze bandage, which is more or less the same thing). Stick this across the inside of a joint with oodles of PVA, and your building will never fall apart.

Adhesives for Card

PVA is, these days, the principal adhesive for card, and jolly good it is too - much better than Croid, Seccotine and all those other noxious trouser-ruining potions we used to use. All PVA is not the same, though - it comes in a variety of strengths and formulations, and there's little to be gained in using expensive PVA woodworking glues like Resin W or

A fine example of card structure modelling using Howard Scenics brickwork - Vincent de Bode's model of Eye station, Suffolk, for his 4mm. 'Flintfield' layout..

Unibond for card work; basic craft PVA is miles cheaper, and works just as well. I get mine in bulk (5-litre drum, about £4.50) from a local educational suppliers, but many art material suppliers like Crookes Craft of Sheffield (who do all this sort of stuff by mail-order) sell it in more modest quantities, usually 1 litre squeeze-bottles.

Other useful adhesives are the various contact cements like UHU, Bostik 1, Evo-Stik and the like. Some of these are pretty noxious on the fume front - maybe not such a pong as Seccotine, but worse for you, one suspects! Best of these I've found so far is the new 'Eco-UHU' (non petroleum based) sold in the Netherlands and Germany, and now available here. I find this new adhesive both easy to use and highly effective on card - either alone, or in combination with a wide range of other materials including plastics.

Paints and Sealants for Card

One of the joys of card is that - provided that you haven't got glue on the surface - it takes water-based paints, including artist's watercolours, so well. I've consistently found I get on better with water-based paint when painting structures - hobby acrylic, gouache, acrylic gouache and artist watercolours. The last, as used at Pendon, are capable of very subtle and delicate effects.

Protecting paintwork and card both from handling damage and the ingress of moisture is an important aspect of finishing card models. Acrylic paints are, in themselves, quite a good sealer, as is ordinary emulsion paint. I quite often give structures an overall coat of white emulsion as an undercoat/sealer if I'm going to paint them with acrylics, not forgetting the back of the card/inside of the model and any cut or exposed edges.

If I'm using watercolour on raw card, then matt artist's watercolour fixative is the answer, brushed liberally onto the surface once the paint is dry. Once again, it's important to paint or seal the inside of the model and those exposed edges - paint them liberally with acrylics. Another useful - and very traditional - sealer is Shellac, sold these days as knotting or button polish. This is a meths-based varnish that seals and stiffens card - postcard soaked in shellac cuts very cleanly and, once painted with acrylic, can represent gloss-painted wooden or metal surfaces very nicely.

Basic Tools for Working Card

The card-modellers toolkit is probably the most straightforward of the lot: A good, clearly marked steel rule for measuring and as a cutting straightedge, a sharp pencil (not too soft - I use an H), a small square of some sort (school geometry-set type quite OK), a cutting mat - A3 size for preference - and, of course, a sharp craft knife or scalpel. I still favour Swann-Morton products, either the old brass-handle craft knife or a No. 3 scalpel handle, always with a curved blade - No. 2 for the knife, No. 10 for the scalpel. A heavier Stanley knife is also useful for cutting up thick card. The only other essential tool is a scriber with the tip rounded (or a bull-nose darning needle gripped in a pin-vice). All these necessities can be obtained from general tool merchants such as Squires, Fourtrack or Eileen, but for one-stop card-modelling shopping, including paints, fixatives, PVA and so on, the specialists are:

Crookes Crafts, 33 Pickmere Road, Sheffield, South Yorkshire, S10 1GY Phone 01142 668198/685339 Fax. (Also shop - 94a Crookes, Sheffield).

Freestone Model Accessories (card kit specialist), 28 Newland Mill, Witney, OIxon, OX8 6HH. Phone: 01993 775979

Dave Doe's exquisite Dutch farmstead also uses Howard Scenics 4mm. embossed card to represent grote steenen - big bricks - in HO scale on his P87 Dutch period layout, 'Portiershaven'.

AYE TO EYE

Chris Archbold F.B.C.O. takes a look at your eyes and gives a few tips on the maintenance of the most important item in your toolkit.

Perhaps it's the way I look - gullible and an easy pushover? - but there I was innocently looking around the Watford Finescale Exhibition when I suddenly found myself face to face with Mike Peascod. 'How's the new magazine going?' I asked and then immediately bit my tongue. A doctor never asks after the health of one of his patients whom he happens to meet in the street, as it leaves him wide open and defenceless.

'Great' was the reply and then: 'But you know, old Ricey - always had excellent vision, saw distant and near objects exceedingly well, boasted of his eyesight. Well, for some weeks now, he's let on that near objects are not as sharp as before, and now he's found out he needs reading glasses. So we reckoned you might do an article on modeller's eyesight for a future edition.'

I have never been hit with a verbal sledgehammer before. My immediate thoughts were, 'I have just bought the preview issue of a new magazine hoping to find out more about this immensely varied hobby of ours and instead am being asked to write the dratted thing. What do I know that everybody else didn't find out twenty years ago?' However, the way Mike put it sounded like a compliment and a challenge - so I accepted.

The Symptoms
So, you're in your mid-forties or over and wondering why you are having difficulty seeing the detail on the models you are trying to make and thinking that if only you had longer arms everything would be alright. 'If only I could see it comfortably I would make a much neater job,' you say. Of course, a possible solution is a change upward in gauge to 7mm. or even Gauge 1, but then how does one squeeze that dream layout of Clapham Junction into a three bedroom semi when one is modelling in 7mm.? If this is your situation, don't worry, help is at hand in the shape of several visual aids that are available.

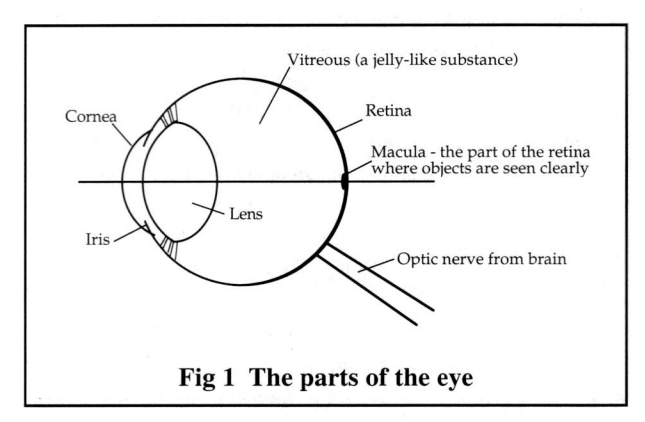

But before rushing into solutions let us first consider what is happening to cause the problem. You are in all probability experiencing the first signs of a condition known as presbyopia, which - although it sounds like it - is not a disease, but is entirely age-related. Parallel rays of light from a distant object are bent or refracted by the *cornea* (the transparent window of the eye) and the *lens*, which is about the size and shape of a Smartie only transparent and sits just behind the *iris* (the coloured part), to a focus on the *retina*. *Fig. 1* is a very basic diagram of the human eye just so you can see where all the 'bits' are.

On a perfect eye, this happens without any effort or strain being involved, but in order to focus on an object nearer than infinity when light rays are still diverging from that object, then an alteration has to take place in the eye. What happens is that the lens thickens and thus changes shape, so that its curvature is greater. This curvature increases the nearer the object is, until a limit is reached and very close objects are blurred no matter how hard one strains.

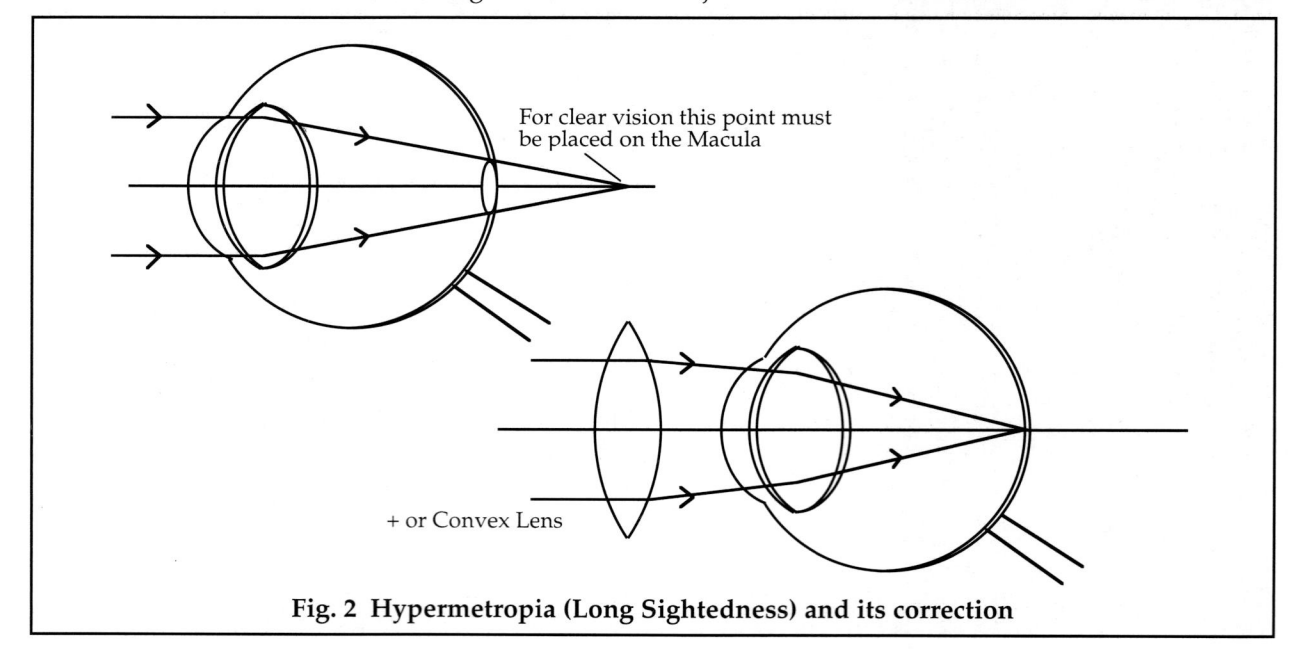

Fig 1 The parts of the eye

Fig. 2 Hypermetropia (Long Sightedness) and its correction

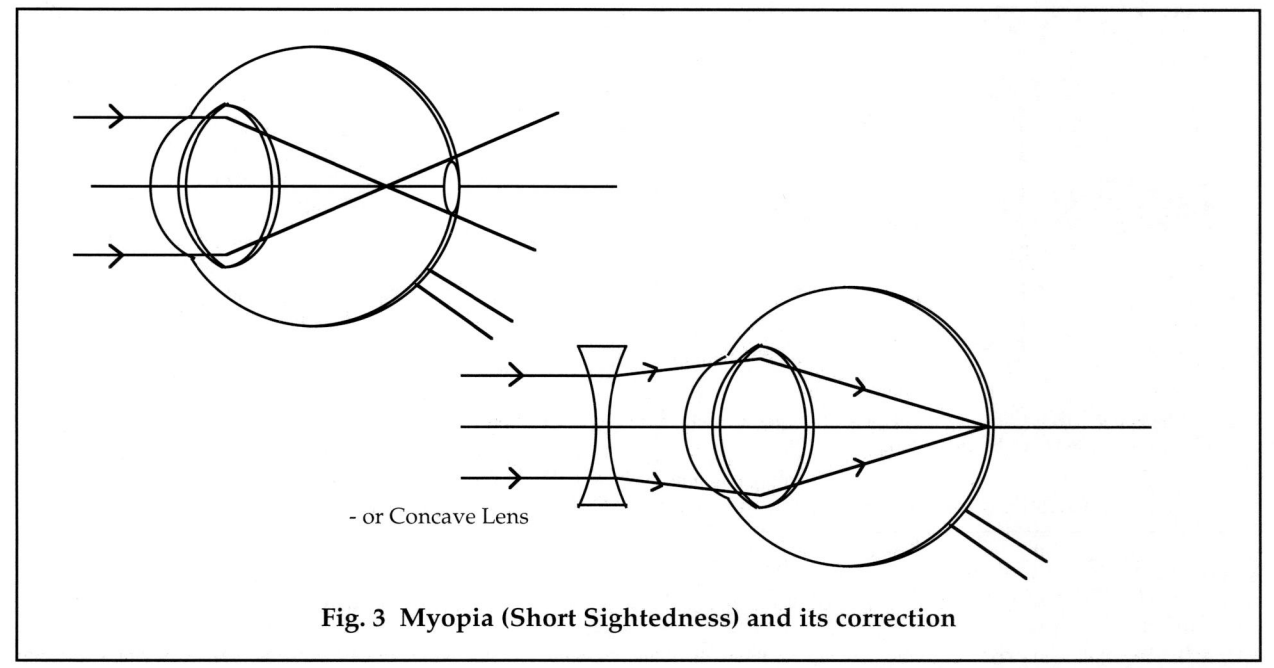

Fig. 3 Myopia (Short Sightedness) and its correction

When young, the eye is very flexible and very close things *can* be viewed with ease. As age creeps on (gallops is probably more like it!) this flexibility diminishes until the limit of strain, or 'accommodation' as it is known, is at a distance further than the objects one wishes to see - and so problems arise. Initially, you find that if you hold the objects at a greater distance then it is easier to see them, and it is understandable to think one is suffering from *asthenopia* or, as it is technically known, eye-strain.

Eye Mechanics
Let us now look at the mechanism governing all this. Without boring you too much with a treatise on physical optics, a few brief words on the three main types of refractive errors seem in order.

Long sight or *hypermetropia* is where the focal length of the eye is beyond infinity and is corrected with a convex or positive lens (prefixed with a +) *See fig 2.* (Light rays are diagrammatically shown with little arrows to indicate their direction, unlike the patient I once saw who was convinced he emitted radiation from his eyes!)

Short sight or *myopia* is where the focal length of the eye is, you've guessed it, closer than infinity and is corrected with a concave or negative lens (-). *See fig. 3.*

The third refractive error is *astigmatism* and is a mixture of the above two present in the same eye. To put it more simply in most cases the cornea or window at the front of the eye is misshapen and is similar in shape to an egg. This egg shape can be horizontal, vertical or at any one of 180 degrees. It is corrected using a cylindrical lens where the power is in one direction only and is often combined with a spherical lens. *(fig 4.)* We need not concern ourselves too much with this as those of you with any significant amount of astigmatism are probably wearing a correction already.

There is another type of lens occasionally used for the correction of muscle imbalances, where both eyes do not look straight ahead together without effort and binocular single vision may need assistance. These are prismatic lenses, and have no 'power' as such - but possess the property of being able to 'bend' light to make it appear to come from another direction. We shall see later that these lenses, when used correctly, can be of great assistance to us.

Lenses are classified in power units known as dioptres. A 1.00D lens will have a focal length of 1 metre. The focal lengths of other powers are obtained by taking the reciprocal of this. i.e. A lens of 2.00D will have a focal length of 50cms. (100/2 =50) and so on.

The Solution
Armed with all this knowledge we can now proceed to try and improve the vision of our ageing nay, maturing modeller. Several solutions are possible, and one man's meat may well be another's poison - so one should be prepared to experiment until a satisfactory and visually comfortable answer is found. This sounds expensive, but with a little assistance it needn't be.

1) A pair of spectacles. These are made up to individual prescription and can be in the form of: basic reading glasses, (with probably a separate pair for distance vision as well); bifocal glasses, where there is a distance portion at the top of the lens and a section at the bottom for near work - this can take the form of anything from a small circle to a straight line running all the way across the frame; or, thirdly, varifocals, where there is no line separating the two portions but they are blended very cleverly in order to give an intermediate section as well. These latter lenses give the most natural effect - but it must be remembered that the reading section is quite small, smaller than on any bifocal, and in the early stages of acclimatising to these glasses one may need exaggerated movements of the head to get everything in focus. Once mastered, they are probably the best way of putting back the clock to a lost youth!

2) A magnifying glass. This is the simplest solution of all. It consists of a simple spherical convex lens and is handheld or mounted on a stand and

Light rays in a horizontal plane are focused to a point but light rays in a vertical plane are unaffected by the lens.

Fig. 4 Refraction by a cylindrical lens

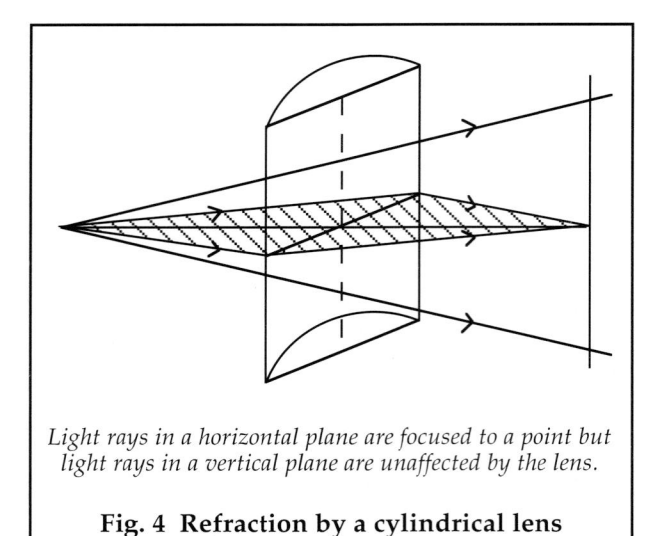

positioned a short distance away from the object to be viewed. They are available in a variety of powers, and the higher the power, the greater the magnification - but also the greater the 'aberration' or peripheral distortion. This aberration increases dramatically as the power, or curvature, of the lens rises so that a powerful magnifier is not necessarily the best.

3) Loupe. A loupe is a magnifying or convex lens worn on a head-band with the lenses held a short distance in front of the eyes - the commonest make of these in modelling circles is probably the 'Optivisor'. In order to see objects clearly one usually has to position the head a short distance from the object to be viewed. These aids are quite effective, but can be tiring to use for long periods.

4) My own preferred solution. A Bishop-Harmann magnifier. This is only applicable if one is in the age group of approx 40-60yrs with distance visual acuity of 6/9 or better. i.e. A car number plate can be seen easily and clearly at 25 yds without glasses. A Bishop-Harmann magnifier takes the form of a very small pair of Dickensian round eye spectacles. They are worn on the end of the nose and have a power of +5.50 Dioptres in each eye coupled with a prism of 4 prism dioptres to assist in convergence. To enlarge on all this very sudden technical terminology, a lens situated near the eye

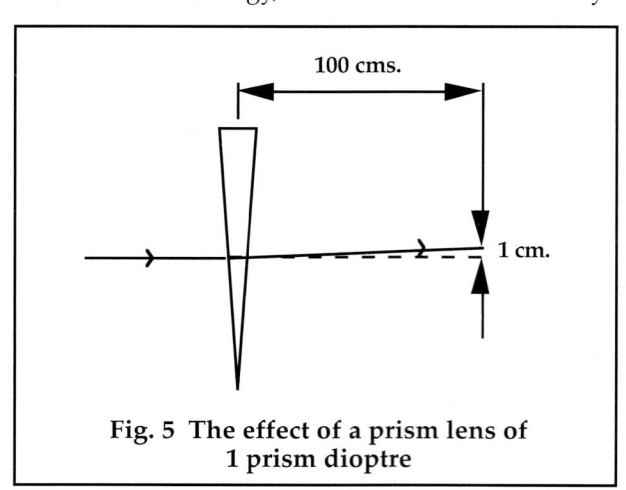

Fig. 5 The effect of a prism lens of 1 prism dioptre

possesses different properties to one held near the viewed object. For example, if an object is at a distance of 20cms from the eye it follows that an accommodative power of 5.00 Dioptres (100/20=5) is needed to see it clearly. If one is only able to exert an accommodation of, say, 2.00D, then a lens of a further 3.00D is required to make up the shortfall. Elementary, my dear Watson...

However, things are not always as simple as they appear because, having corrected for this distance with a +3.00D lens, the eyes are still having to converge from 'straight ahead' to meet at 20cms. distance - a convergence of 5.00 prism dioptres. (One prism dioptre is the lens power required to move an object visually by one centimetre at a distance of one metre - fig 5 should make this clear.) One's powers of convergence also decrease with age - although not nearly so rapidly as that of accommodation. But in order to obtain visual comfort, a prism of 1-1 1/2 prism dioptres with the base nasally (i.e. fat bit on your neb, thinner bit at the top) may well help. A prism is a flat sided lens with no 'power' as such, but light entering one side is refracted so that it emerges in a different direction.

Lighting
'I can see it first thing in the morning or in sunlight.' Yes, you can when your eyes are fresh and the light is good. After a day's straining in poor light, things aren't so clever. Good lighting is as important an aid to good vision as any artificial aid. It is a fact that as we become older the sensitivity of the retina diminishes, so that at the age of forty twice as much light is required to see clearly as at age twenty. This deterioration is progressive, so that at sixty, three times the amount of light is required and so on.

Bright light also causes the pupil to contract and the smaller the pupil, the less diffusion of light entering the eye and thus the sharper the image seen. This can be demonstrated by making a small pinhole in a piece of paper and then holding this up to the eye and viewing an object through it. Using this method one can often read quite small type without the aid of reading glasses, but obviously it's not a very practical solution!

There are various options available nowadays to improve things. About 150 years ago there was only candlepower (but there were not many railways to model either!) Gas lighting was introduced a century or so since, and was a big improvement, but not until the invention and introduction on a large scale of the electric light was it possible to illuminate intricate, close work to enable it to be done at night with ease. Nowadays, of course, with modern fluorescent lighting, there is no problem.

Workshop Lighting
In the workshop, one needs to set up a good background light level. This can be achieved by using a fluorescent tube, or preferably two. But extra illumination will still be required on the workbench, and this can take the form of 'Anglepoise' lamps or similar directable fittings. I know the label on the shade of the average Anglepoise advises a maximum wattage of 60, and that 60 watts does not give a very high level of illumination; but reflector bulbs such

as those used in shop window displays (the PAR 38 range of Fresnell reflector floodlights or the R60/R80 reflector spots, all available in 60, 75 and 100W outputs) are now available with a bayonet fitting, and these direct all the light forward onto the work area. I use two lamps with this type of bulb, one wall mounted and one on the side of the workbench. If you want to go one better then there are florescent tube Anglepoise lamps which can give up to the equivalent of 100 watts.

(It is also possible to 'uprate' an Anglepoise lamp by improving the cooling; drill additional cooling holes about 5mm. in diameter in the upper part of the shade, and paint the shade itself matt black to improve radiation. You can also replace the plastic lampholder with a porcelain equivalent, and in this form you can put 100W or even a 150W PAR 38 bulb in and really light the place up - I.A.R.)

Summary
I hope that at least a little of the foregoing is understandable and that you now have a rudimentary idea of what happens to the physical and refractory side of our eyes as we age. Pathological changes are beyond the scope of this article, as are some of the methods used to alleviate the difficulties thus encountered.

What happens later? Well, as the lens in the eye slowly loses its accommodative powers, then stronger and stronger lenses will be required. However, this is a very slow process and takes decades to occur. The question I am frequently asked by worried patients is 'Will this deterioration eventually lead to blindness?' The answer is a very emphatic NO. Refractive changes in the eye do *not* result in blindness. Most people think of blindness as the lights being suddenly turned off - and this just does not happen.

Taking Action
You may well have recognised your own visual difficulties in the conditions I've described. So what should you do next? The answer is simple. Go to an optometrist and have a thorough eye examination. There is simply no substitute for this, as he is trained to spot any possible pathological conditions (i.e. diseases), as well as help you with any straightforward visual problems. I have lost count of the number of people I have seen who had no idea they had anything wrong with them at all - and and turned out to be suffering from, say, high blood pressure.

Tell the optometrist about your hobby, and that you wish to be able to hand-line a 4mm. coach perfectly at a distance of, say, 14cms (would that I could!) - and thus make sure that you get the visual aids you need for comfort. Why not take a model with you and explain your problems - but bear in mind that, whilst trying to help, miracles are a little more difficult, even for optometrists. It is just not possible to have a lens which will enable you to see a sparrow's kneecaps at a hundred yards *and* paint a spotted tie on a model figure in 2mm. scale two inches beyond the end of your nose!

In conclusion, if you feel that a lot of close work will injure your eyes, let me hasten to set your mind at rest. Comparative observation has shown that well-lit close work does not cause any injury to the eye so long as an adequate correction is worn if necessary.

I hope that all this waffle has not been too technical and difficult to follow - and that you are still wide enough awake to read the rest of the magazine!

Chris Archbold is an Optometrist with practices in Bedford and Martock, Somerset. He is also nutty about railways, especially the Great Northern and LNER.

The Australian Journal of Railway Modelling, aka 'Branchline Modeller'

Our reciprocal arrangement with Stephen Ottoway and ProtoHO is at long last bearing fruit, and by the time this appears we shall have all issues of Australia's premiere fine scale modelling magazine in stock.

So - if you fancy a change of diet, want to be the most cosmopolitan member of your club, or just have a yen to poach a few ideas from the bright folks down under (and they do seem to have more than a few bright ideas) - then you know what to do.

Copies of **Branchline Modeller 1 - 4** and **Issue 1 of Australian Journal of Railway Modelling** are available at £4.75 each + 75p. P & P from:

RailMODEL, PO BOX 2 Chagford Devon TQ13 TZ
Phone/Fax: 01647 433611
Visa/Mastercard welcome

Issue No. 4

4mm. Wagon Running Gear
Sprung Suspension Systems

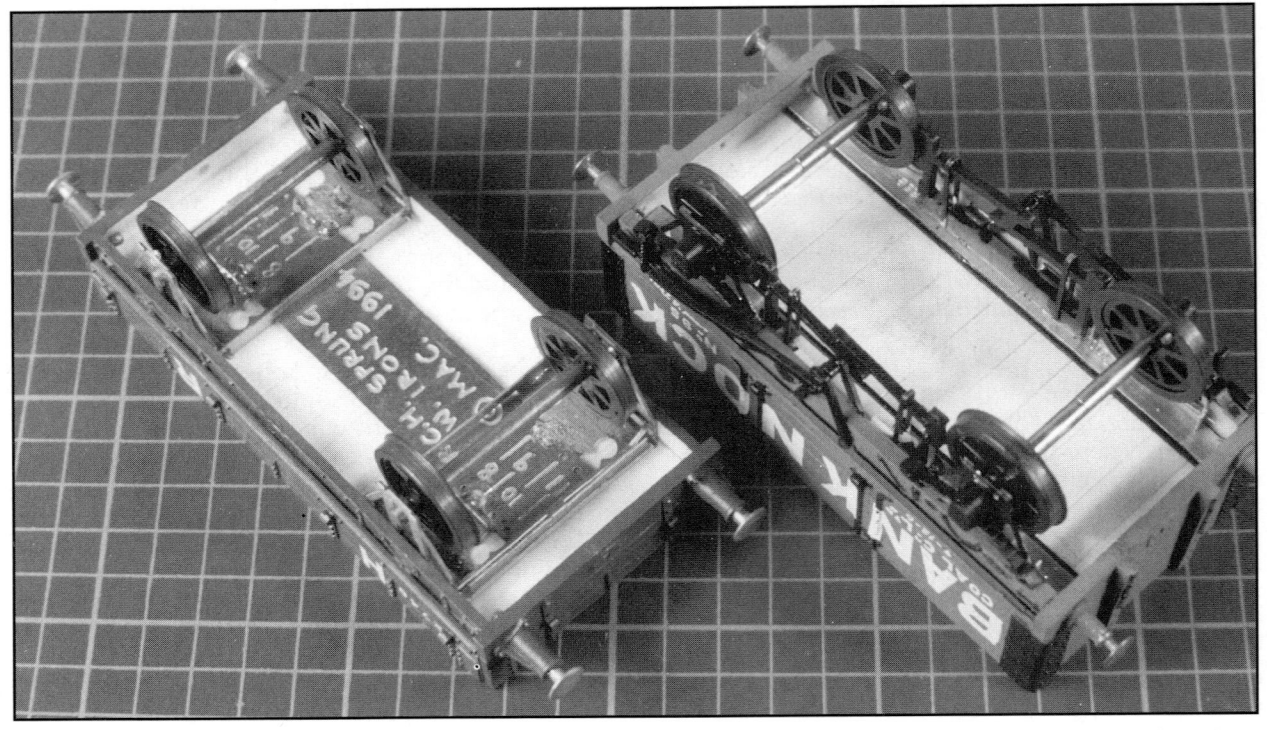

The rival systems for 4mm. sprung wagon system - Masokits on the left and the Exactoscale alternative on the right,

4mm. *After his examination of the various 3-point compensated wagon suspension options for 4mm. scale wagons in* **Digest 3**, *prolific wagon-builder* **Simon de Souza** *gets to grips with the sprung alternatives, comparing systems from Exactoscale and Masokits.*

Although there were isolated examples of 4mm. rolling stock springing systems in the past - such as the ERG internal springing units or SRB's working sprung axleboxes - it is only in the past few years, with the introduction of systems from Exactoscale and Masokits, that springing has become a standard type of fitting for 4mm. freight stock. These two makers approach the task from different angles and are worth considering separately.

Exactoscale
The Exactoscale springing units were first on the market and come in two forms. The first, basically developed for fitting to kits, has the 'W-irons' or axleguards joined in pairs along each side with a 'dummy solebar', fixing the wheelbase. This is available in lengths of 9ft., 9ft. 6ins. and 10ft. The second is a 'custom fit' version, in which each axleguard is completely separate, very much like the prototype. Indeed, fidelity to the prototype is one of the most striking features of the Exactoscale system.

The basic system consists of a tiny axlebox slider (described as a 'bearing cradle') unit which slides in a hornway in an etched-brass W-iron of more-or-less conventional design. The actual spring is of fine spring-steel wire; it is retained to the axlebox by a tab and bears against a spring-seat on the W-iron or dummy solebar etch. The action of this spring is very soft, and excellent riding results from even modestly-weighted wagons, so long as the axlebox unit is truly free to slide in the W-iron.

The wheelsets need new axles to replace the standard 26mm. pinpoints. These are made up of two 2mm. diameter brass sleeves fitted over a fine steel inner axle - the ends of which protrude and which then run in tiny (really tiny!) parallel bearings. A jig is supplied to help make the axle - which is assembled with Loctite - go together, but I suspect for some people this may be a feature which puts them off. Certainly, care is needed to ensure good running, with the wheels true and in gauge. Comment has been made about the increased rolling resistance of the parallel bearings - but I suspect that unless you run very long trains indeed, then this is a problem which has been exaggerated.

Installation
Assembling these units is not without other difficulties. I found the main problem to be the steel-wire spring, which is little thicker than a human hair, and which fits under a fold-up tab on the sliding inner axlebox. This is *very* easy to lose and rather fiddly to fit. Indeed, given the immense amount of thought and care which Bernard Weller always puts into his products, I found this part of the assembly rather awkward and perhaps susceptible of some improvement. I have also broken several of these fold-down tabs off in the past in trying to fit the spring, but perhaps I'm just heavy handed.

The other awkward feature in using the individual axleguard units is the actual business of assembling the axleguards to the solebars; despite the provision of a plethora of jigs, in effect, you have to deal with four very small, separate parts, all with adhesive applied, plus the wagon, plus the various jig components, simultaneously - not easy! Bernard does suggest in his copious instructions that, with a brass wagon, you could apply solder cream, jig the whole thing up, and sit it on a hot plate to be soldered in one fell swoop. However this is obviously only an answer when fitting to all-metal wagons that have been assembled with a relatively high melting point solder - not that common a scenario these days. Resistance soldering might be a better option.

Performance

Once assembled however, the results from these units are very good and they do have a number of very real plus points - not the least of which is greatly reduced lateral movement of the wagon body, by virtue of the parallel bearings. All other compensation types have at least one axle running in coned pin point bearings and these do allow some lateral movement. The reduction of this slop with the Exactoscale units is of particular benefit to those using Alex Jackson couplings.

Another considerable advantage to the Exactoscale range is the wide variety of different prototype axleguards (W-iron patterns and axleboxes) that Bernard has produced and, I believe, continues to introduce. For some vehicles with distinctive axleguards, these units may well be your only choice. Most other makes, either of compensated or sprung units, only supply the obvious RCH/GWR standard type plus the common BR and BR 'plate' axleguards, although D&S do produce a wider choice in compensated units. However, unless you are extremely 'picky', then the RCH/GWR W-iron will do perfectly well for most steam-era vehicles - but if absolute accuracy is needed then the Exactoscale product is 'de rigeur'.

Summary

In my experience the fixed-wheelbase 'kit conversion' version of this product is considerably easier to use, so if you want to experiment my advice would be to try a set of these first; a set fitted my 9ft. WB Slaters PO wagon very well. As the commonest wheelbase lengths are covered, you will probably be able to equip most of your scratchbuilt wagons with these as well as kits.

Once installed, the spring and sliding cradle are, to all intents and purposes, invisible. Axlebox and spring castings can be fixed 'rigid' to the outside of the W-iron etching, with only minimal clearance needed for the (tiny) floating bearings.

The Exactoscale system calls for modified wheelsets on non-pinpoint axles, made up of the inner steel axle plus brass sleeves seen here; the standard wheels are refitted to the new axle.

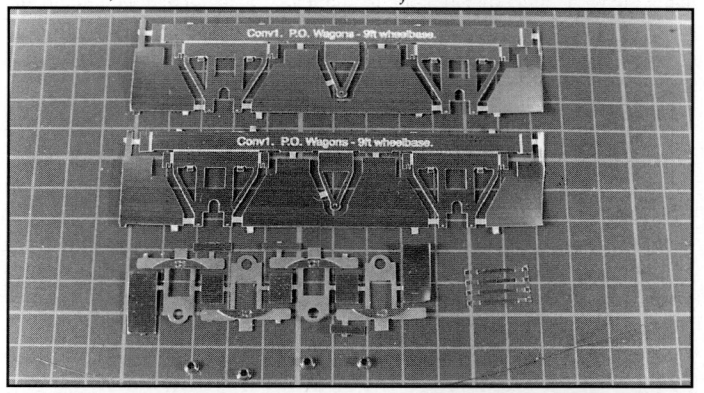

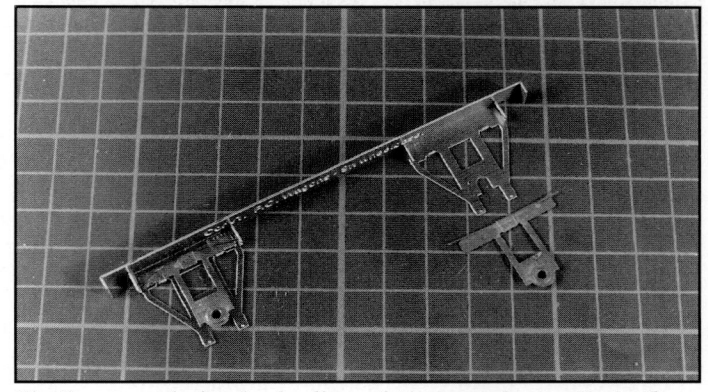

The Exactoscale fixed-wheelbase 'kit conversion' units in the flat (top) showing the paired W-irons on dummy inner solebar, plus the 'bearing cradles' that carry the fine-wire springs and act as sliding axleboxes. The whole system is shown installed at the left-hand W-iron in the lower photo.

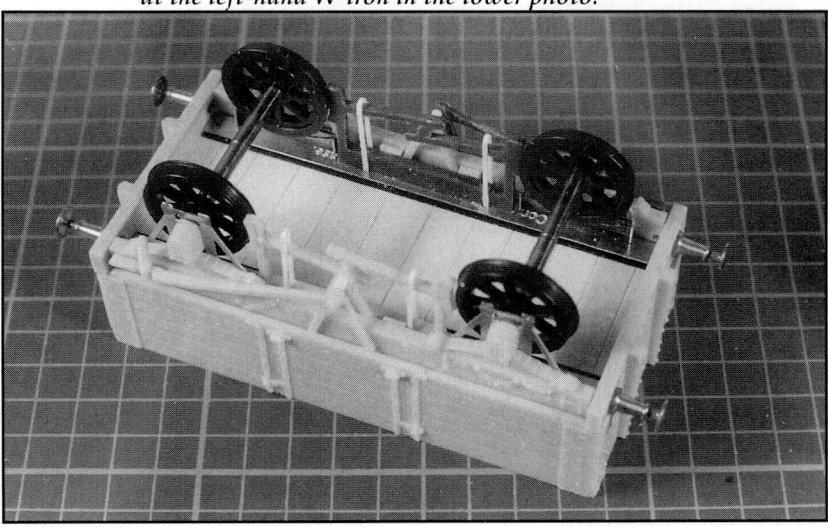

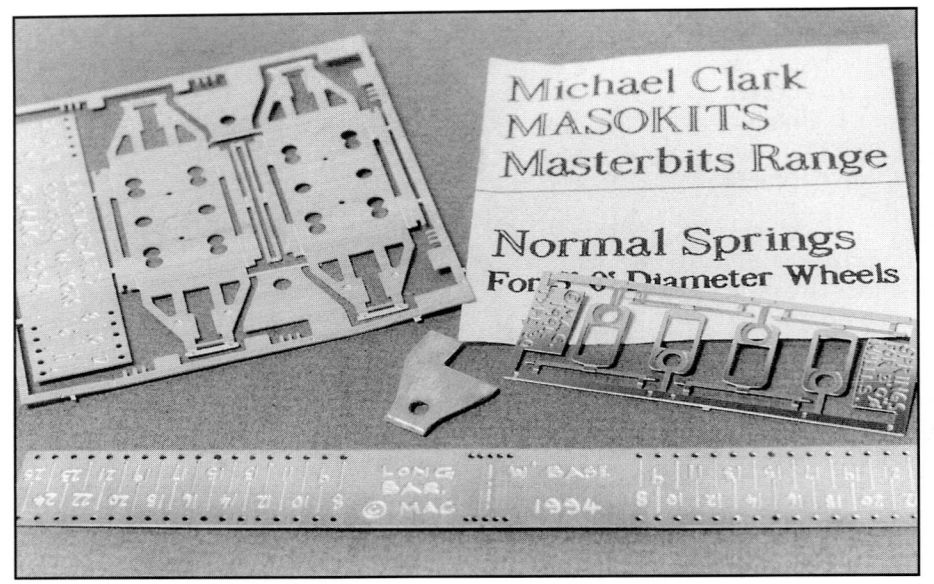

The components of the Masokits system - etched 'paired' W-irons, bearing carrier/springs in stainless steel, wheelbase spacer 'spine' plate.

Masokits

The Masokits springing units, produced by Mike Clark, at first sight more nearly resemble 'traditional' compensation units. They retain the normal pinpoint bearings (with the associated potential lateral movement mentioned) and have a thin leaf spring, made integral with the inner (moving) axlebox and produced by folding the half-etched top part over by 90 degrees (Bill Bedford uses a similar technique in some of his wagon kits). These inner units are, unusually, etched in stainless steel - which results in a very flexible spring that, while not quite as soft as the Exactoscale wire, gives a good springing action.

Assembly and Installation

The Masokits W-irons fold up into full width units which are then pinned with brass wire to a 'spine' bar running most of the length of the wagon beneath the floor, and having location holes at a scale 6-in interval; once final locations are established, the units are soldered to this spine, which is simply stuck beneath the wagon floor - it's easy to pack for a correct ride height. This is a simple mounting system and in this regard, the units are self-jigging for parallel.

The stainless spring unit/bearing carrier is folded up, and the normal top-hat pinpoint bearing retained in place with cyano, making sure to keep this away from the sides of the bearing which slide in the slot in the W-iron. There should, however, be no fore-and-aft slop in the fit of the bearing in this slot. I also found it a good idea to file the tops of these bearings where they came through the W-iron back a bit, to ease the fitting of cast axleboxes later. As described by John Chambers in his review of the Masokits system in *Digest 1*, you can solder this stainless steel, using Carr's Brown Label flux and normal 145º solder.

Axlebox/spring castings used with this system do need the holes in the axlebox opening up a bit to allow the bearings to move - although you don't need much 'travel' for the suspension to work effectively. I suppose it would also be possible to separate the axlebox from the spring and fix the box to the bearing so that it moved with the wheelset in prototypical manner. As with the Exactoscale units care is needed to avoid paint, stray glue, solder or whatever gumming up the working.

Summary

Results for both types are excellent although I must confess I found the converted axles of the Exactoscale type gave me some trouble in obtaining smooth running and the 'custom' type, particularly, more complex and fiddly to put together than the Masokits version. Both manufacturers score particularly high marks for the quality of their instructions - no skimping here - and the ride of the sprung wagons is noticeably smoother than compensated stock, partly because one slight disadvantage of any three point compensation is the bump when the fixed axle meets an irregularity in the track. This last quality may not necessarily be to everyone's taste; Chris Challis has pointed out that the traditional British non-fitted goods wagon actually didn't run very smoothly in practice!

Weighting

One crucial aspect of this whole wagon undergear business is that of weighting. It matters not whether you build your wagons solid, compensated or sprung, without sufficient weight you will not obtain good running. The received wisdom is says 20-25 grams per axle and once again I refer you to the manuals of the EM Gauge Society and Scalefour Society or to Iain Rice's books 'Detailing and Improving Ready to Run Wagons' and 'Getting the Best from Plastic Wagon Kits' for further details.

How you get the required amount of lead (roof flashing bought off the roll at your local professional hardware store) into your wagon can require a good deal of ingenuity. A small flat wagon with steel channel solebars (only 3mm. deep rather than the 4mm of a timber underframe) will not have enough space underneath for sufficient lead - you'll need to fit a load made of whitemetal! On the other hand, a van is - of course - simplicity itself and if your wagons run loaded at all times (no empties on your railway - the prototype companies *would* have admired your efficiency!) you likewise have no problem.

Otherwise it is a matter of cramming as much lead as possible into every nook and cranny underneath. Dr. Weller rightly concludes that his springing units, with no components between the solebars, leave more space for ballast, although the gain is only very slight - and you can always put some lead inside the compensating units themselves. Bear in mind also that with sprung units that the ride height will alter depending on how much weight is fitted; for the Exactoscale units, it may be necessary to change the thickness of the spring wire on heavy vehicles. Bernard Weller gives full chapter and verse in his copious instruction book.

Conclusions

Some problematic prototypes clearly lend themselves to one form of suspension rather than another. In particular, for most people, tank wagons and also some modern hopper types, with their very open underframes, make conventional three point compensation virtually impossible - there is just nowhere to hide the compensating units. Thus, for those who do not want to leave the underframe solid, Exactoscale springing is a Godsend. Having said which, Geoff Kent has come up with some very ingenious home-made compensation units, described in his excellent book 'The 4mm Wagon - Part 2 ' (Wild Swan Publications). Part 1 is also required reading for aspiring 4mm. wagon-builders, I would suggest.

However, where such specialised problems do not intrude, the choice as to which of these many suspension systems (and I would not claim the list to be exhaustive) to use is really, I believe, down to ease of use and the personal preferences of the modeller. Any of the methods I have outlined will, given care in construction, produce excellent results. Most of the many wagons I have built have been compensated and I would still suggest that, for most people, this is the simplest and certainly the quickest way to improve the running of a wagon, be it RTR, kit or scratchbuilt.

Of the compensated options, the screw-thread mounted type is perhaps the soundest from the engineering viewpoint but requires a little more work from the builder. The straightforward tab and slot is stunningly simple to use, as long as you are using good quality components such as MJT units, but I have also been very impressed by the rubber band mounted vehicle I built for the purposes of this article and I intend to experiment further with this.

A jig to ensure that axles are parallel

It is of course essential that wagon axles, as well as being square to the longitudinal axis of the wagon, are also exactly parallel with each other. The simple Plastikard jig in the photo takes literally three minutes to make and will easily ensure that this desirable state of affairs prevails. The photos are largely self-explanatory: The main base of the jig being a piece of 40thou. sheet about the length of the wagon, and of a width to fit between the wheels (in P4, I made this 17mm).

Three more pieces of 40thou. - preferably cut from the same strip as this baseplate - are then produced, the longer middle one being the length of the wagon wheelbase minus 2mm - the diameter of the wagon axle (ie., allowing half-a-diameter at each end). Thus, for a 9ft. wheelbase wagon in 4mm. scale this piece should be 34mm. long. It's obviously vital that the ends of these spacer pieces are square to the length and truly parallel - an engineer's square used with reasonable care ensures that the jig is accurate in this respect.

One of the short pieces is glued in place at one end with solvent and checked with the square. The axle of a wheelset is then located against this and the longer wheelbase-spacer piece butted up hard against the axle. This too is glued in place with solvent, and finally the process is repeated with the second end piece. When set hard, the jig is used as in the second photo - holding the axles in parallel while the adhesive securing W-irons or springing units in place on the wagon cures. Other jigs for different wheelbases are easily produced in the same way.

'S' Scale rocking solebar compensation.

One form of compensation I omitted from my survey in Part One was the method advocated for very many years by the 'S' Scale Society. This keeps the axleguards in unit with the solebars, one of which is fixed as normal. However, the other solebar is arranged with a pivot halfway along its length, rocking about this point to give the three-point effect.

Personally, I can't see any great advantage over any of the other methods described, and feel it is perhaps less easy to set up than most 3-point types - although if you find yourself cornered by any of the 'S' scale gang they will argue very persuasively for it! The Society can supply etched axleguards produced for them by Bill Bedford; unless he will produce them for 4mm. scale, I am not aware of any similar 4mm components, although I suppose it would be possible to use 'old-fashioned' one piece whitemetal axleguards. If you want chapter and verse my advise would be to join the 'S' Scale boys and girls and thereby get hold of their excellent and highly entertaining manual!

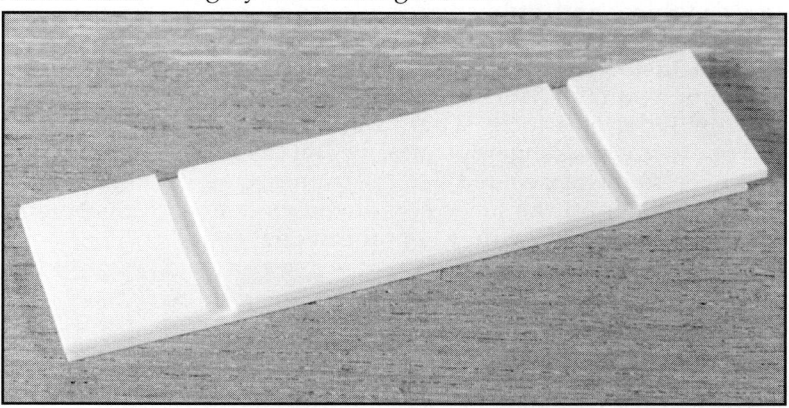

D-I-Y Plastikard Wagon axle spacing jig

LANKY SWANSONG -
THE LAST TWENTY HUGHES 4-6-0 EXPRESS LOCOMOTIVES.

Prototype Study

Noel Coates and Peter Priestley prepare a modeller's guide to a striking class of locomotives, giving convincing reasons for their appearance on many LMS layouts.

10455: *As running in the early 1930s with the 12in. gold numerals, shaded vermillion and lake, and the 'S' plate below the cab roof.*
Photo N.G. Coates Collection

The engines which are the subject of Peter's drawing, were built at Horwich between April 1924 and January 1925 and feature detail differences from their forebears, differences which made them appear more balanced engines. The main purpose of this article is to explain for modellers the differences found upon individual engines within this last 20, the liveries they wore and where they were likely to be seen. Thus hopefully satisfying the main query of those interested: Can I justify running one on my layout?

Introduction

These engines were originally ordered as 4-6-4 tank engines, and published evidence suggests that the frames were already in preparation when the conclusion was reached that there would not be enough suitable work around for such powerful 4-cylinder *tank* locos. There was still a shortfall of express passenger locomotives on the LMS, however, so it was decided that the frames should be shortened and the locos completed as tender engines instead. Thus, they kept the deeper framing and longer wheelbase bogie which had been developed for the front end of the Hughes L & Y Baltic tanks (11110-9), which gave 10455 - 74 a more mature look compared with earlier 4-6-0's. On completion, all were immediately drafted onto the West Coast main line to fill traffic demands working the Crewe to Carlisle section. Many spent the greater proportion of their lives working over Shap and, although the Royal Scots very quickly took over the most important trains, these last 20 Hughes 4-6-0s held on to the secondary expresses for a further five or so years (1928-33), until the Patriots grew in numbers and displaced them.

10456: *As a compound at Carlisle Upperby in the early 30s with the 14in. gold/black numerals; shows the cab roof profile to best advantage.*
Photo N.G. Coates Collection

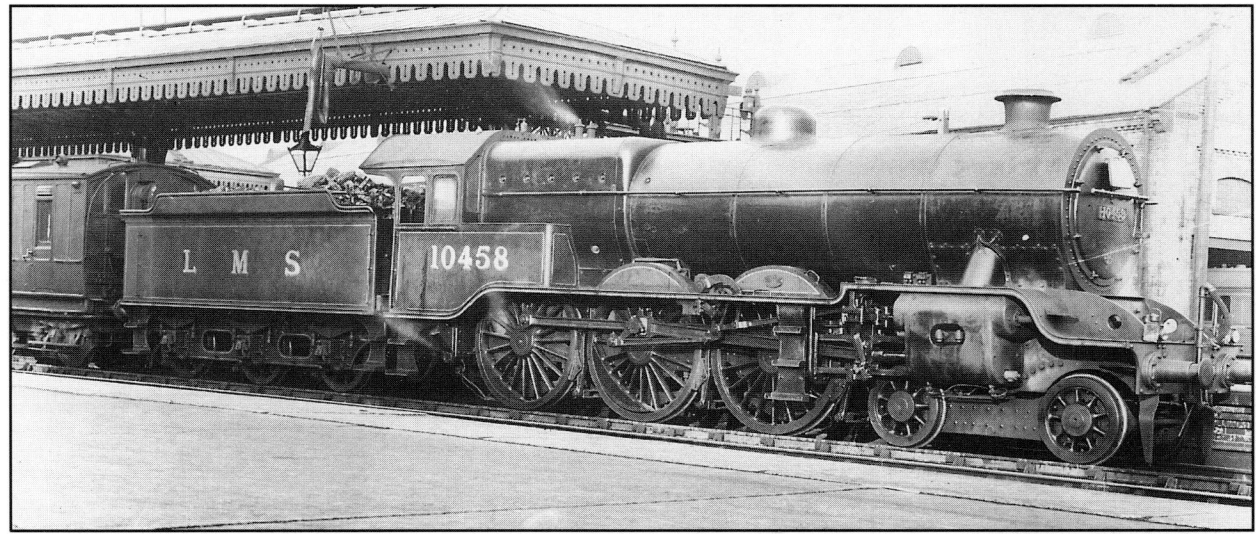

10458 : Sunlight glinting on the firebox shows some of the livery to best advantage, though a lot of the interest is up front with the wider Kylala chimney, rivets on the smokebox front. new style shedplate 24E: (Blackpool) and no lubricators; about mid 1935?
Photo N.G. Coates Collection

The engines then drifted onto the Central Division where they forced some cascading - but the 'scrap and build' policy of 1933 caught them cold and, as their boilers became due for renewal, they were withdrawn and replaced by new Stanier 4-6-0s. Within 25 months (February 1935 to February 1937) 17 of them had been scrapped, the survivors being 10455/60/4; a very short life indeed for engines which had once been thought of as the ultimate in express motive power.

The Hughes 4-6-0 Family
The Hughes lineage went back over 15 years to the emergence of his first 4-6-0s (highly imperfect in several ways), and these final 20 were the fifth variant. Nor did the development stop there; the first alteration was the conversion of three of the last 20 - Nos. 10459/62/74 - and 16 of the earlier engines to oil burning in late spring 1926, a result of the prolonged miners' strike, which included the General Strike. The oil apparatus was not in use for very long, but the cylindrical tank on the tender was both obtrusive and crude - there was no visible external alteration to the locomotive itself.

The next was, perhaps, the most interesting development, involving the conversion of 10456 to a 4-cylinder compound, which emerged in July 1926. The basic idea was to gain data for the compound 4-6-2 and 2-8-2 designs proposed by the new CME,

Fowler, but the loco was hardly ever tested. The only detail differences were the very wide steam pipes, as the external cylinders were identical, together with the removal of the lubrication box from above the right hand cylinder; the engine spent most of its time working between Crewe and Carlisle and was much liked by enginemen.

In 1927 at least seven of the 20 were upgraded to first-class mechanical condition, allowing them to work on top link duties; this was shown by a prominent letter 'S' after the cab side windows just below the roof. 10455/7/60/7/9/70/2 were the members of this exclusive club; the letters seem to have been retained when the locos were allocated to the Central Division in the early 1930s.

The final development involved engine No. 10458, which emerged in mid-1928 with a slightly wider chimney (and different lubrication again), though it was within the smokebox that the major alteration had taken place with the fitting of a Kylala blastpipe. Like all the devices tried on the Hughes engines it was only a marginal improvement and hardly changed the look of the engine - though it is significant for the modeller. We have tried to illustrate all these minor alterations.

The last survivors received a wonderful ring of rivets around the outside edge of the smokebox faceplate from late 1935 or thereabouts. One other detail difference which must be mentioned is that

10465 : As built and turned out in dark grey with 3in. yellow numerals below the cab windows and no evidence of ownership.
Photo N.G. Coates Collection

10473 : *The right hand side of a normal engine with 14in. gold/black numerals - the commonest arrangement carried by 13 out of 20 engines.* **Photo: Real Photos**

4-6-0 engines could be lifted using the front buffer beam and pivoted on the rear driving wheels quite happily; this was not so on a 4-6-4T, where the rear bogie would have suffered immense strain. So, while the engines already laid down as Baltics had lifting-holes in the frames, the frames of the last 11 were cut as 4-6-0s and thus had no need for these holes.

Liveries

Although the locomotives qualified for the red livery they were first turned out in a dull dark grey with yellow 3in. serif numerals below the cabside windows, and it was to be some time before they were painted red. The exception was 10474 which, having been chosen as a LMS representative for the Stockton & Darlington centenary, was given a special finish - complete pre-1928 livery and full insignia with 18in. numerals. As the engine was in traffic in February 1925 and the parade was not until July 2nd it must have run in the dark grey or something similar for a few weeks.

However, it seems to have been several years, about late 1927, before the other 19 finally began to receive the red livery due to them, presumably after a general repair or boiler change (e.g. 10455 first swapped boilers in February 1928, getting the one which had been on 10469) an ideal time for full livery application. Examination of photographs has revealed that some engines hardly altered whilst others were positively chameleon-like where the numerals were concerned. Perhaps the easiest way to explain this to describe the findings for each locomotive:-

10455 - no direct evidence until the early 1930s when it carried 12in. gold shaded red (actually vermilion to right and lake below) numerals. By the late 1930s it had received the 12in. chrome yellow shaded red numerals (of 1937). It gained its BR number and black livery with LNWR style lining in October 1948.

10456 - (as a compound) - it carried 12in. gold shaded black numerals up to 1930 or so, thereafter it had the 14in. gold/black type until withdrawal.

10457 - the only evidence shows 14in. gold/black style after the chimney been altered - but a picture taken shortly before withdrawal shows the 12in. gold/red style.

10459 - no evidence has been discovered.

10460 - this engine was the chameleon, it seems that every time visited the works it came out with something different - carrying, in order: 12in. gold/black, 14in. gold/black, 12in. yellow/red numerals and finally, as a wartime black repaint, (the only loco to receive this) the 12in. yellow/red again (dates of changes not known).

10461 - a 1932 photograph shows the 14in. gold/black style.

10462 - listed as carrying firstly 12in. gold/black numerals and latterly 14in. gold/black numerals

10463 - a July 1934 photograph shows the 14in. gold/black style.

10464 - a cornucopia of numerals here and one of only three engines known to have carried the 10in. gold/black numerals; by May 1935 it had 12in. gold/black changing to the 12in. yellow/red in the late 30s.

10465 - photographed about 1930 with the 14in. gold/black numerals.

10466 - listed as having the 14in. gold/black numerals.

10467 - listed as having the 12in. gold/black numerals.

10468 - photographed in August 1933 with the 14in. gold/black numerals.

10469 - listed as having the 14in. gold/black numerals.

10470 - photographed late 1934/early 1935 with 12in. gold/red numerals.

10471 - photographed on Horwich works in September 1936 awaiting breaking, still with 10in. gold/black numerals.

10472 - photographed in July 1933 with 10in. gold/black numerals.

10473 - mid 1930s photographs reveal 14in. gold/black numerals.

10474 - the full pre 1928 treatment (detailed above) and retained whilst an oil burner, latterly the engine carried the 14in. gold/black numerals.

10472 : At Crewe in July 1933 and still with the 10in. gold/black numerals and 'S' plates. Photo :Gordon Coltas

The photographs included in this article reveal some of the liveries described oposite. So even for a small class of 20 engines, there is much variety to choose from. When you consider that they had occasional trips to Glasgow and that an odd excursion took them to London (10470 in Summer 1934), there is scope for the modeller whose interests lie outside the West Coast main line north of Crewe or the LMS Central Division to utilise one of these magnificent red locos.

References and further reading.

The L&Y in the 20th Century; *Eric Mason* (1954, Rev.1961), Ian Allan

Locomotives Illustrated No. 88, (1993), RAS Publishing, ISSN 0307 1804.

West Coast 4-6-0s at Work; *C.P. Atkins* (1981), Ian Allan, ISBN 0 7110 1159 1.

Illustrated History of LMS Locomotives Vol. 2; *Essery & Jenkinson* (1985), OPC, ISBN 0 86093 264 8. (Vol.1, 1981, ISBN 0 86093 087 4, tells more in general about engine colours, lining & lettering)

British Locomotive Catalogue 1825-1923, Vol 3B L&Y Railway; *D. Baxter (ed.)* (1982), Moorland, ISBN 0 903485 0.

10474 : In super-finish mode at Farnley Junction shed on the way north for the centenary celebrations in late June 1925.
Photo: A.G. Ellis Collection

50455: Left and above ~
Two views of the last survivor, at home (Blackpool) and at Manchester Victoria on the last run (useful rear view of lubricators).

Photos, both
N.G. Coates Collection

Below ~
A detailed view of the right hand gear.

Photo
N.G. Coates Collection

Summary of details.

	Built	Wdn	Age Yrs	Mnths	Notes
10455	7/24	10/51	27	3	S
10456	7/24	3/36	11	8	Compound from 7/26
10457	7/24	8/36	12	1	S
10458	8/24	11/35	11	3	Kylala from mid 1928
10459	8/24	2/36	11	6	Oil burner 1926
10460	8/24	12/47	23	4	S
10461	9/24	5/35	10	8	
10462	9/24	10/35	11	1	Oil burner 1926
10463	9/24	9/35	11	0	
10464	10/24	6/39	14	8	
10465	10/24	1/37	12	3	
10466	10/24	8/35	10	10	
10467	11/24	1/37	12	2	S
10468	11/24	2/36	11	3	
10469	12/24	10/35	10	10	S
10470	12/24	8/35	10	8	S
10471	1/25	2/36	11	1	
10472	1/25	2/35	10	1	S 1st withdrawal/shortest life.
10473	1/25	2/37	12	1	
10474	2/25	12/35	10	10	Exhibited 1925, oil burner 1926

S = Tip top mechanical condition

10460 : *Close up view of the left hand valve gear and reversing rod, in November 1934.*

Photo N.G. Coates Collection

Oil burner : *Not the right engine type but the correct tender for the last 20, showing the single oil tank (not a pretty sight but saves worrying about modelling the coal!). L&Y No. 1656, later 10427, in 1926; the Attock carriage behind is already lettered LMS!*

Photo A.G. Ellis Collection

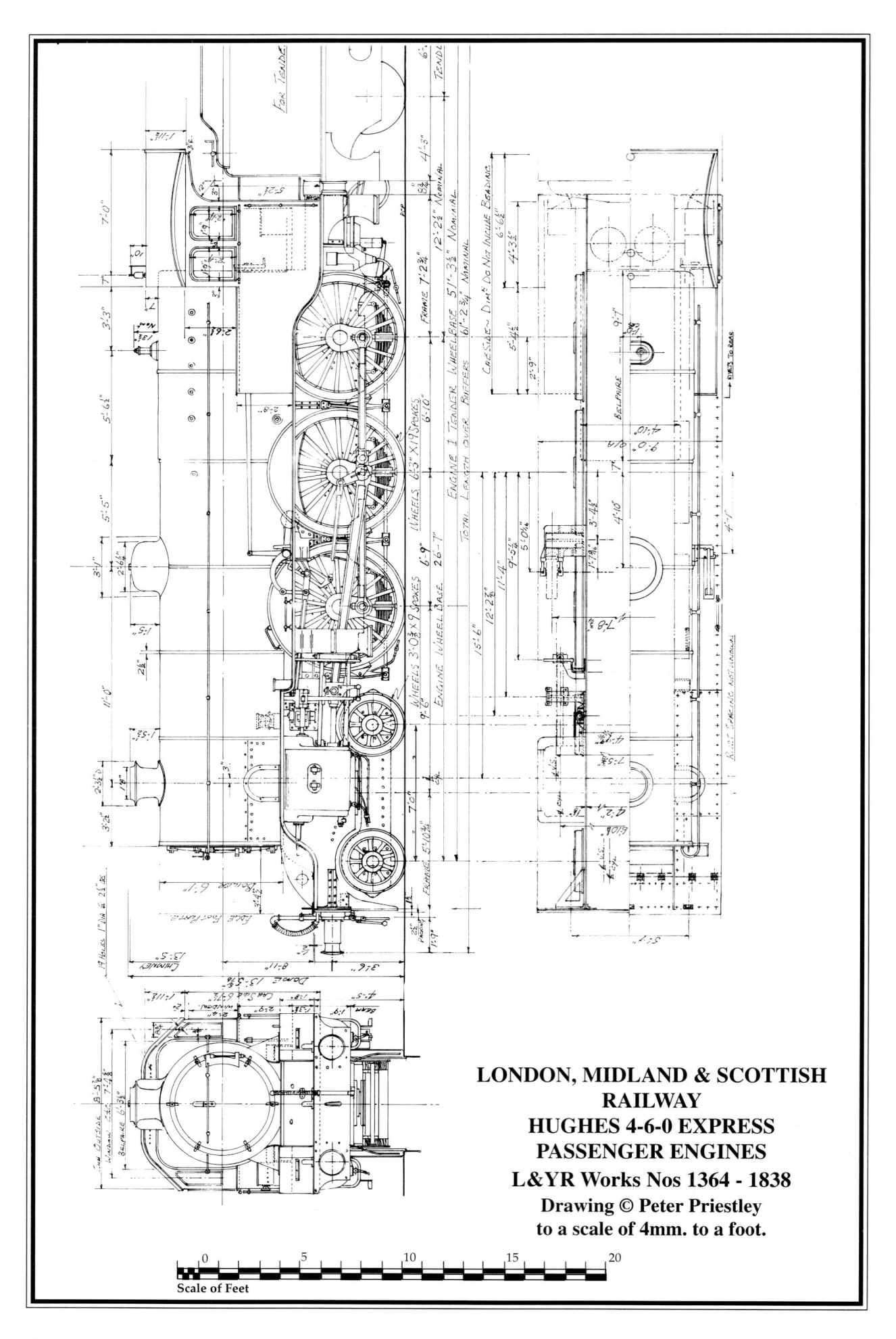

LONDON, MIDLAND & SCOTTISH RAILWAY
HUGHES 4-6-0 EXPRESS PASSENGER ENGINES
L&YR Works Nos 1364 - 1838
Drawing © Peter Priestley
to a scale of 4mm. to a foot.

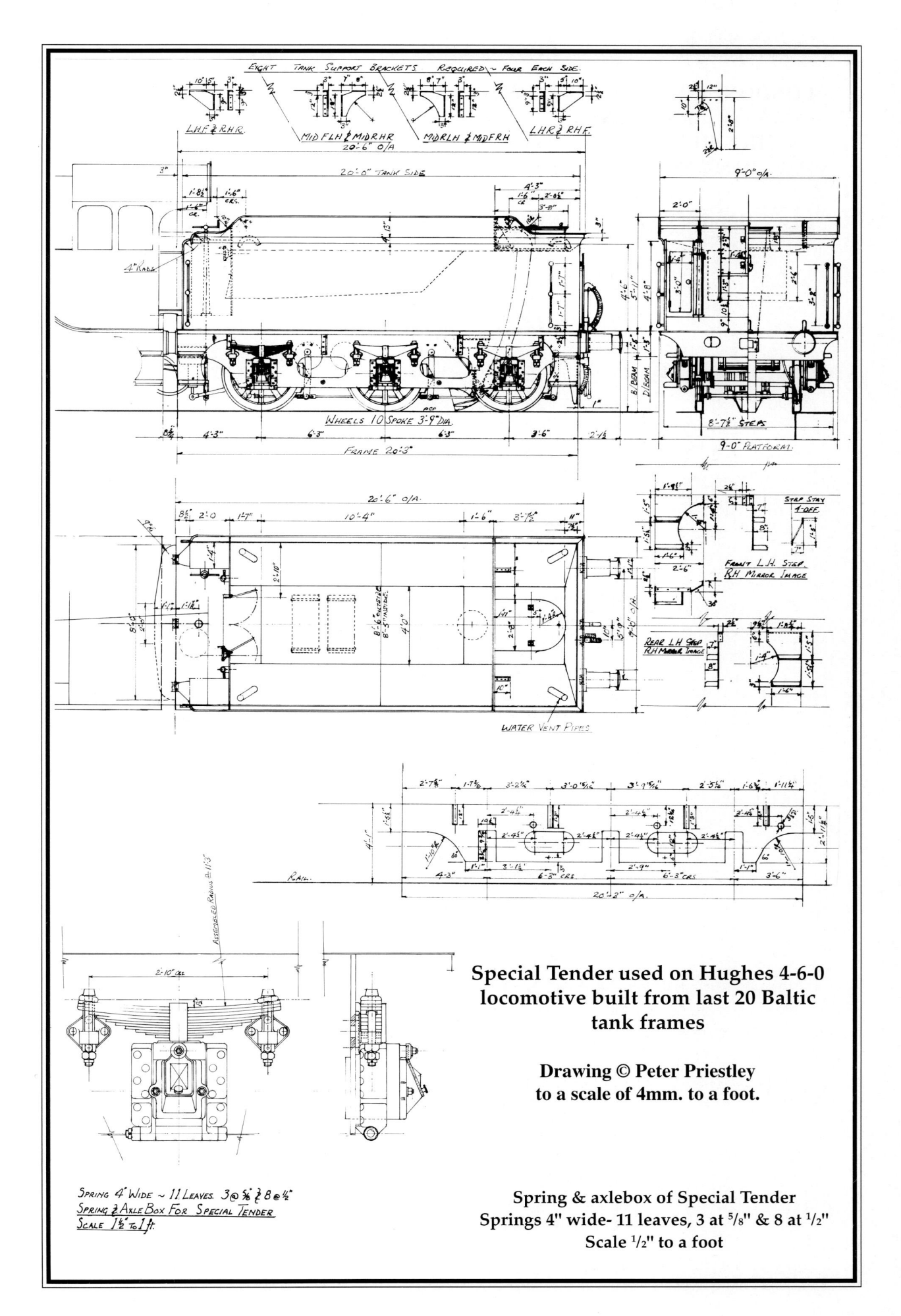

Special Tender used on Hughes 4-6-0 locomotive built from last 20 Baltic tank frames

Drawing © Peter Priestley
to a scale of 4mm. to a foot.

Spring & axlebox of Special Tender
Springs 4" wide- 11 leaves, 3 at $5/8$" & 8 at $1/2$"
Scale $1/2$" to a foot

LONDON, MIDLAND & SCOTTISH RAILWAY
HUGHES 4-6-0 EXPRESS PASSENGER ENGINES
Details
Drawing © Peter Priestley

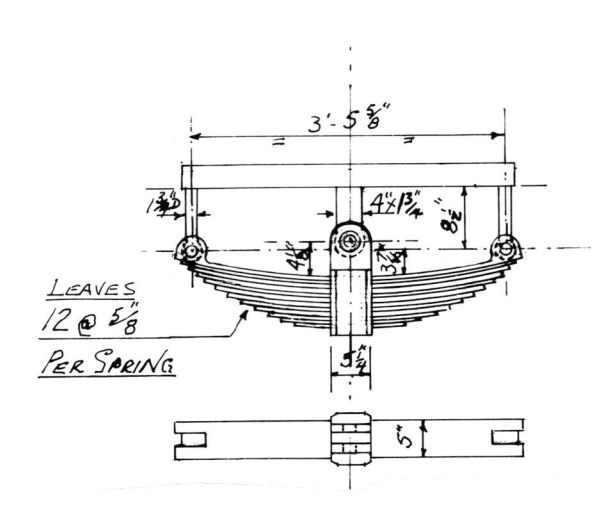

Detail of loco spring
Scale 1mm. to 1 inch.

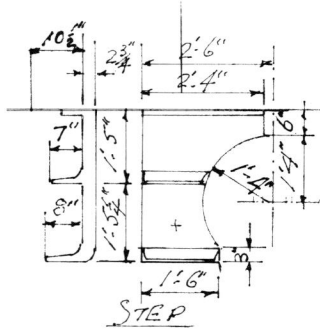

Steps - Left hand shown, right hand mirror image.

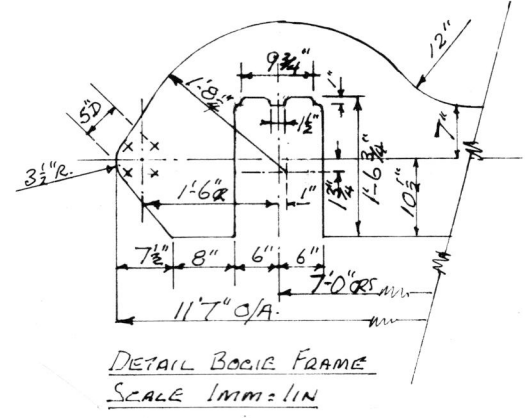

Detail Bogie Frame
Scale 1mm. to 1 inch.

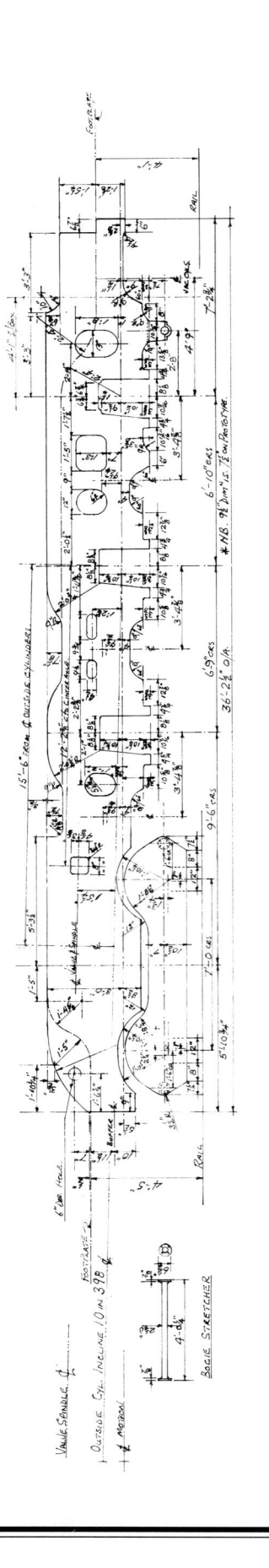

Prototype Matters: Cattle Wagons

The transport of livestock - especially cattle - by rail was a traffic which developed from the earliest years to a peak - probably some time about World War 1 - then steadily declined, although not in spectacular fashion until the advent of the motor cattle lorry in the 1950's saw it off. As with so many British wagon designs, the cattle wagon evolved only slowly, as these two examples - built all but half-a-century apart - show; in many respects, they are startlingly close: Same capacity (8 tons), wheelbase and overall length., same type of running gear. Only the steel underframe and vacuum brake of the BR-built wagon are a significant advance on the 1906 Midland wagon. But then - there's more than a ton added to the tare weight; real progress? Actually, the BR wagons saw very little use in the traffic for which they were built, as they arrived at the same time as the 10-ton 4-wheel lorry; much of their working lives (a lot shorter anyway than their Midland ancestors) was spent in carrying fruit, ale and other non-bovine traffics.

    The other distinction that these two cattle wagons have is of being probably more popular in plastic-kit model form than their prototype was. The MR Large - in its early LMS form -was the subject of the very first 'scale' plastic wagon kit of all, George Slater's original body-only mouldings introduced back in the mid-1950's, when all other offerings in the genre were freelance. It subsequently, in a very fine new version, became an early staple of the modern Slater's wagon kit range. The BR van, of course, was the subject of one of the mould-breaking and very significant Airfix two-bob kits back in about 1962 or 3. Given a little chunkiness about its (working) hinges, this is a good kit that can still hack it in 1996. A longer life - by far - than the protype. The example photographed below is the NRM's preserved example; cattlevan in aspic?

Photos:

Top: BR Official
Lower: Iain Rice.

Under the LENS...
3mm. on Parade

3mm. scale, being a little out of the mainsteam of railway modelling, tends to attract modellers with that streak of individuality and independence - which can make for some very attractive models of less-obvious subjects. Here are a selection, as seen at the 1996 AGM of the 3mm. Society and photographed by Geoff Helliwell.

As well as being the 3mm. Society's ex-officio snapper, Geoff Helliwell is no mean loco builder, as this scratchbuilt GW 'Metro Tank' testifies. This is a neat, wormanlike model of a neat, workmanlike engine, mercifully devoid of any of the copper-capped, brassbound fripperies that so many GW modellers seem unable to resist applying to their locos. 1498 looks as though she has seen plenty of service - a very nicely detailed portrait of a working locomotive. It won the Tony Birch Trophy at the 1996 3mm. Society AGM

3mm. is a scenic-modellers scale par excellence, and the 'off line' classes of the AGM competition always attract a good entry. Well out of the ordinary is Nick Salzman's thrashing set hard at work, with scratchbuilt thrashing box and elevator. The engine - a small Fowler? - is a modified toy. Very nostalgic and nicely observed.

Peter Bossom models the SR in wartime Britain, an unusual subject in any scale! This 2-NOL electric set is a rarely-modelled piece of classic SR first-generation 3rd. rail stock, built from Roxey etched parts and finished in 1930's olive livery with varnished droplights. This delightful model captures the distinctive character of these SR sets very well, and repaid close study for a well-modelled interior, a subtle weathering job and some very neat glazing - really thin, flush and accurate. The model deservedly won the Cuckmere Trophy at the 1996 3mm. Society AGM.

All photos: Geoff Helliwell

The Croydon club have been building a model of the Culm Valley line in 14.2mm. gauge fine scale for some time, and as part of the project Mike Davey built this model of the complex dairy plant that provides the main source of traffic for the line. Those open windows are special etchings and are all hinged; the doors open, too, while the brickwork is hand-cut, hand laid chads! The dairy chimney stack - only the base shows in this shot - is a real eye-opener, and looks most convincing. This was another 1996 AGM competition success - 2nd. place in the Ralph Murfitt Trophy.

Pre-Group GLORY

Our subject this issue is the LNWR water troughs at Bushey in Hertfordshire, with a pair of F. Moore paintings showing two generations of LNWR express trains taking water at this famous location. The first of these dates from around 1892 -3, and shows one of the infamous Webb uncoupled 2-2-2-0 3-cylinder compounds of the 'Teutonic' class, built in 1889, on the 'Wild Irishman' (Sic! - presumably a nickname for the Irish Mail; one can't conceive of Sir Richard Moon entertaining such a title officially!) The huge 27-inch low-pressure cylinder - driving the leading axle - is clearly visible between the frames. The stock of this crack express, leaving aside the vans at both ends of the train, is six-wheeled - although at least of uniform roof profile and outline - not always the case with LNWR express trains at this period.
Oilette Postcard

Same location, but some fifteen years on, an unidentified express heads north over the troughs behind a new Whale 'Precursor' 4-4-0, a simple modern design compared with the eccentric Webb compounds they were built to replace. Dating from 1904, these workmanlike 4-4-0's ushered in a new era on the LNW, the first of a family of sturdy if unexceptional engines that shared only names with their famous predecessors. Although the steel-framed tenders of these Whale engines looked modern, they still had the archaic design of horseshoe tank with coal well, which restricted water capacity somewhat. The train is now of bogie stock, the 57ft. design with the three-arc roof, modern and comfortable with corridors and all mod. cons.
Loco Publishing Co. Postcard.

BLAGDON, S & D Jt. Section.

Low sun emphasises the detail on Ivatt 2-6-2T 41241 (71G Bath GP, outshedded Radstock), the Blagdon Branch loco this week. The model is, of course, based on the excellent Bachmann RTR version, detailed and subtly weathered. As with a lot of current-generation RTR, it's wonderful what an improvement in realism results from a little toning-down of shiny wheel-rims, valve gear and the like, plus the fitment of scale (or at least less-obtrusive) couplings.

All photos: Neil Burgess

41241 draws to a stand with the 10.50 a.m. ex-Binegar. In the background, the vegetable patch (essential component of any rural S & D station) is newly-dug. The goods lock-up is a grounded van body - reputed to be of S & D origin, and the end-loading dock is occupied by a PW engineer's mess and tool van (once a Midland tarrif brake) and a brand-new vacuum-braked ballast hopper.

Above: The passenger facilities at Blagdon. The station building is a rendition of the smaller type of Bath Extension structure - essentially Wellow in mirror-image. The parcels lock-up is from Radstock, while the nameboard and lamps hail from Binegar. The footpath to Blagdon village climbs up from the rear of the platform. The gradient post on the platform end indicates the recommencement of the 1 in 65 up towards the reservoir; as was usual practice where possible, the platform road has been kept level.

Most of Blagdon can be seen from this view from above the goods yard end, with some rather full-sized daisies in the background! The open gate onto the Blagdon - Burrington road at the end of the loading dock serves as the usual pedestrian access. The line to the reservoir carries on past the water tank (from Sturminster Newton, less the 'bag'). The two-coach branch train is composed of Airfix LMS suburban stock, the leading lavatory compo being as original, but the brake third having acquired a set of Comet period 1 etched overlay sides.

The BLAGDON BRANCH
S & DJ Section

Johnson 1P 0-4-4& 58051 has run round the branch train and now stands ready to depart on the run back to Binegar. The loco is built from a Craftsman kit, and the coach is a detailed Airfix model.

4mm. The believable might-have-been has long been one of the most effective layout-design themes open to the railway modeller seeking authenticity without necessarily wishing to don the relative strait-jacket that often follows a decision to replicate some actual location faithfully.

Very far removed from the traditional 'free for all' often called to mind by the term 'freelance', this combination of fact with fiction calls for a thoroughly-thought-out rationale, something that **Neil Burgess** has been to some pains to establish for his delightful 'Blagdon Branch'...

Introduction

It is a bright but cold Friday morning in late March 1954 as we alight on the down platform at Binegar station on the Bath extension of the Somerset and Dorset Joint line. The time is just twenty to eleven, and the overnight rain has left large pools of water on the crumbling platform surface - between which we dodge, having alighted from the brown-upholstered splendour of the Maunsell three-coach set which has conveyed us from Bristol, via Bath, in relative warmth and comfort. Walking down the platform we pass the doorway of the cavernous baggage compartment of the leading brake third, from which station staff are unloading boxes onto a barrow, and pause briefly to look into the cab of the locomotive, Bath Shed's recently-acquired ex-LMS '2P' 4-4-0 40527; the fireman is taking advantage of the stop to feed a few rounds into the firebox, and the safety valves are just beginning to lift. The driver looks down, exchanges a few words about the weather. "Had a Tin Lizzie (Bullied light pacific) last week," he comments, as we complain of the March chill "warmer than this 'un."

At the foot of the platform ramp, we pause to let the train depart. With a shriek of the whistle, 40527 blasts her way out of the station, making easy work of the three coaches as the train gathers speed for the last mile or so of climbing to Masbury summit - after which it is mostly downhill and plain sailing to Evercreech Junction and on to Templecombe and Bournemouth. As the brake-end sweeps by, we catch a first view of our next train, the two coaches for Blagdon sitting patiently in the short bay platform on the 'up' side. With no time to look closely - it is almost ten to eleven and departure time is fast approaching - we notice that the train is composed of an odd pair of old ex-LMS coaches, a brake third from the 1920's and an even older Midland elliptical-roofed lavatory composite, hauled by Bath's modern Ivatt 2-6-2T 41241.

Finding an empty third class compartment we get in and then stand, head and shoulders out of the window, awaiting departure time. The guard's whistle blows, the driver responds with a deep hoot from 41241's Caley-style chime, and the short train moves off, gathering speed as it runs alongside the main line for several hundred yards before veering off to the right and westwards towards the Chew Valley and distant Blagdon. We pull up the window and sink back on the seat, watching the sunshine catch the clouds of steam from the engine as they drift past.

41241 is hardly being taxed by this journey, most of which is downhill, and speed soon rises to the line's maximum of 40 m.p.h. Thus far this morning, the loco has brought her coaches up from Radstock, where she is outshedded from Bath for this duty, and has completed two of the day's seven return trips along the branch. Arriving 'up' at Binegar just after ten, she deposited a small gaggle of passengers for the 10.17 Up Bath, then ran round to await what small traffic our Down Bournemouth might have to offer. There seem to be very few passengers going down to Blagdon on this trip, poised as it is between the workers and shoppers on the first two workings and the homeward drift later in the day.

Indeed, the passenger traffic is somewhat sparse altogether and the future of the line doesn't look any too hopeful: already, the minor branches off the Somerset and Dorset system are being pruned back. The line to Wells closed over two years ago, in October 1951, and the Bridgewater line due to go this autumn. (Indeed, since these words were first written in June 1954, it has now closed - on October 4th.). All this is a far cry from the high hopes with which these lines were projected and built in the last century, promoted in the hope of tapping a trade which somehow never actually materialised.

These gloomy reflections are interrupted as the brakes go on for the first stop at Wells Road, almost exactly two-and-a-quarter miles from Binegar. Drifting in across the level crossing over the Bath - Wells main road the train draws up at the short, empty platform with its wooden buildings, opposite the siding and loading bank. The current occupant of the siding is a solitary coal wagon, its sides draped with empty sacks as the coalman pauses from his shovelling to exchange a quip with our driver. Meanwhile, the porter-cum-crossing-keeper has reopened the gates to road traffic and now walks up the platform ramp to relieve the sole passenger alighting here of her ticket, and to take charge of a large bundle of sticks from the guard. We overhear a snatch of talk about runner-beans before, with a flourish of the flag and a sharp hiss of steam, we're on our way.

Historical Antecedents

The countryside through which we are now passing is Mendip upland, although not as bleak and open as at Binegar. This is quarrying and sheep-farming land, wholly removed from the lush dairy country around comfortably ecclesiastical Wells. The Romans mined lead up here almost two thousand yeas ago but it was the quarrying of stone that brought about the construction of this railway - albeit in horse-tramway form - as long ago as the 1830's. By the 1850's the line stretched from Gurney Slade, near Binegar, to the rich lowlands around Blagdon, but its fortunes - and the traffic flow - were irrevocably changed when the 'Bath Extension' of the Somerset and Dorset was built in the 1870's.

By the time the S & D acquired the tramway in 1885 to convert it to a steam railway, the Blagdon end was all-but abandoned, the stone traffic having transferred to exchange facilities at Binegar. The S & D's real goal in acquiring the remnants was as the basis for a route to Weston-Super-Mare, then fast-developing as a seaside resort. In the event, the parlous state of S & D finances at the time meant that the refurbishing of the line and the short extension to the Blagdon terminus was enough to swallow up the meagre funds available.

The Lower Section

Priddy Station, the first block post, is five-and-a-half miles down from Binegar. As our train makes a cautious approach round the curve and beneath the road bridge, we pass the small corrugated-iron goods shed and the two sidings off the end of the passing loop. The small signal box, in LSWR style, is at the end of the platform, close to the stone-built station buildings. Trains *can* cross here, but the loop is 'goods only', so the passenger service is restricted to single-train status. Several passengers are, however, waiting and a similar number alight to pass the time of day with the signalman-porter who collects their tickets. The ritual few moments' gardening discussion with the guard completed, the porter returns to the box, the signals clear, and we're off again.

The Blagdon branch goods shunts at the terminus, with BR slope-sided and riveted mineral wagons (courtesy of Parkside) cheek-by-jowl with an ex-SDJ 6-wheel 'road brake van'. The loco is 'Bulldog' 43258, the S & D version of the MR Johnson 0-6-0 rebuilt with G7H Belpaire boiler.

Right:
Blagdon Station basking in the sun, as placid and deserted as such country stations usually were, with the branch set's compo sporting the tail-light of the next departure.

Below:
A 'Bulldog' brings the branch goods to a stand in the platform road. The double-arm signal - the original stood at the lower end of the Clandown Colliery branch at Radstock - is an S & D rail-post pattern, with the rails bolted directly together rather than spaced in SR fashion. The model uses Code 75 BH and MSE parts.

Charterhouse, the penultimate station on the branch, is eight-and-three-quarter miles and twenty-seven minutes from Binegar. Another modest station, Charterhouse has stone buildings on the platform, three sidings and a signal box. Although a block post, it has no passing loop, and the signal box is only switched-in for one shift, until the daily goods has departed at 2 p.m. Today, only a couple of passengers join and leave the train, but there appear to be a good many parcels and boxes to handle, all loaded promptly onto a barrow. The local postman is emptying the mailbox on the platform, making the gardening forum between porter and guard a three-way affair before the guard flags the train away. 41241 gives a long hoot as the train plunges into the six-hundred-yard tunnel beyond the station.

Because the line is on continuously-falling gradient, a legacy of its tramway origins, our passage through the blackness is nowhere near as sulphurous as on the up journey, when the loco must slog its way up the 1 in 75. Once out into the March daylight, the countryside has changed noticeably. The dour uplands are giving way to a greener, more fertile landscape rich in trees and grazing cattle. In the distance ahead we catch glimpses of open water - Blagdon Reservoir, built in the early years of the century to supply water to Bristol, and an important reason for the line's survival; the material used in its construction came to Blagdon by train and on to the site over a mile-and-a-half extension beyond - the nearest the S & D ever got to Weston-super-Mare. In more recent years, a second and larger reservoir has been built at nearby Chew Magna, and this undoubtedly provided a further 'shot in the arm' to the branch.

Blagdon Terminus
The gradient changes now, from 1 in 75 down to 1 in 65 up, and 41241 starts to work harder. This is the last section into Blagdon station, the new alignment built by the S & D in 1885. A couple of minutes collar work brings us to the short, curving platform of the terminus, with its tiny signal box and stone buildings, similar to those of the main Bath Extension. The train draws up abreast the buildings, and we alight to find that the chill evident on our brief wait at Binegar has given way to milder weather, although there are still signs of heavy overnight rain.

The fireman is already down between loco and train uncoupling, and as we reach the ramp 41241 draws forward, then stands taking water from the column at the platform end while the fireman climbs back up to walk across and insert the train staff, unlocking the ground frame to throw the loop points. Tanks replenished, the engine moves forward to run round the two-coach train, and we note that there seem to be a fair few wagons in the yard awaiting collection by the daily goods; so perhaps there is some hope the line will keep going a while longer.

On the other hand ...

Anyone scratching their head and wondering why they have never noticed the Blagdon branch in any S & D book can reassure themselves that eyesight and memory have not deserted them. There was, of course, a Blagdon branch - but it came the other way, from the west, being built by the Wrington Vale Light Railway, latterly part of the GWR. Had the S & D line existed as I've supposed, of course, the WVLR would never have been built, which would have robbed Somerset of one of its less usual railways.

My Blagdon is an attempt - born out of a sort of hazy dissatisfaction with my existing layout and a desire to try a few less-familiar constructional methods - at a fairly minimal-space layout - 2.2m. x 0.55m., executed to fine scale 16.5mm. gauge standards. The track layout I arrived would appear to be in the 'Llanastr' family - essentially a reversed and very slightly enlarged version of Iain Rice's 'Elan/Llanastr Mawr' (described in his 'Layout Design - Fine Scale in Small Spaces' book) which, in turn, was derived from Rodney Hall's 'Llanastr'.

For Blagdon, the baseboard front takes the form of a concave curve, and I preferred an end-loading dock to the loco shed of the Rice plan or the goods shed of the Hall original. Blagdon also does without a bay platform, which seemed to me to make for too much complexity. The main part of the construction work has been spread over about eight months, although planning and research went on a lot longer!

Baseboards, Track and Control

The layout's foundations are Norman - Barry rather than 11th. century French - using a frame of twin-member 5mm. ply girders spaced with 50mm. x 25mm. timber blocks, assembled with ⅝ ins panel pins and Resin W. The top is also ply, again 5mm., and incorporates the 1 in 65 grade up to the platform from the Binegar end and thence towards the reservoir, which called for some elementary maths, a bit of careful measurement and some well-placed spacers. As the 'prototype' was supposedly a recyled tramway, it seemed only appropriate to build these baseboards from recycled materials - two ply-and-batten doors retrieved from a builder's skip. Costs aside, salvaged timber is more likely than the scrofulous, misshapen knot-holed stuff offered at the local D-I-Y superstore to be dry and stable; it may even have been properly seasoned in the first place!

The two baseboards are of equal size, aligned by brass dowels and held together by over-centre case catches - both items obtained from Red Dog wood-working at Milton Keynes. The result proved to be remarkably light, strong and simple to assemble or dismantle, making moving it about easy; a truly portable layout.

Trackwork

Plain track is SMP code 75BH flexible, mostly with nickel-silver rail but also using phosphor-bronze 'rusty' track for siding-ends. The pointwork is also SMP, from their PCB components. The lack of chairs on the pointwork I find not too obtrusive - I suppose if it really bothered me I'd do something about it!

The points are operated using an adaptation of the wood-strip-and-rubber band method described by Chris Lammacraft in MRJ 44 - he was good enough to demonstrate it to me at Railwells a couple of years ago. Point crossing polarities are changed by simple separate SPDT switches mounted alongside the point controls but not connected to the operating linkage - remembering to change both point and polarity isn't too much of a problem on a small layout like this.

Ballasting

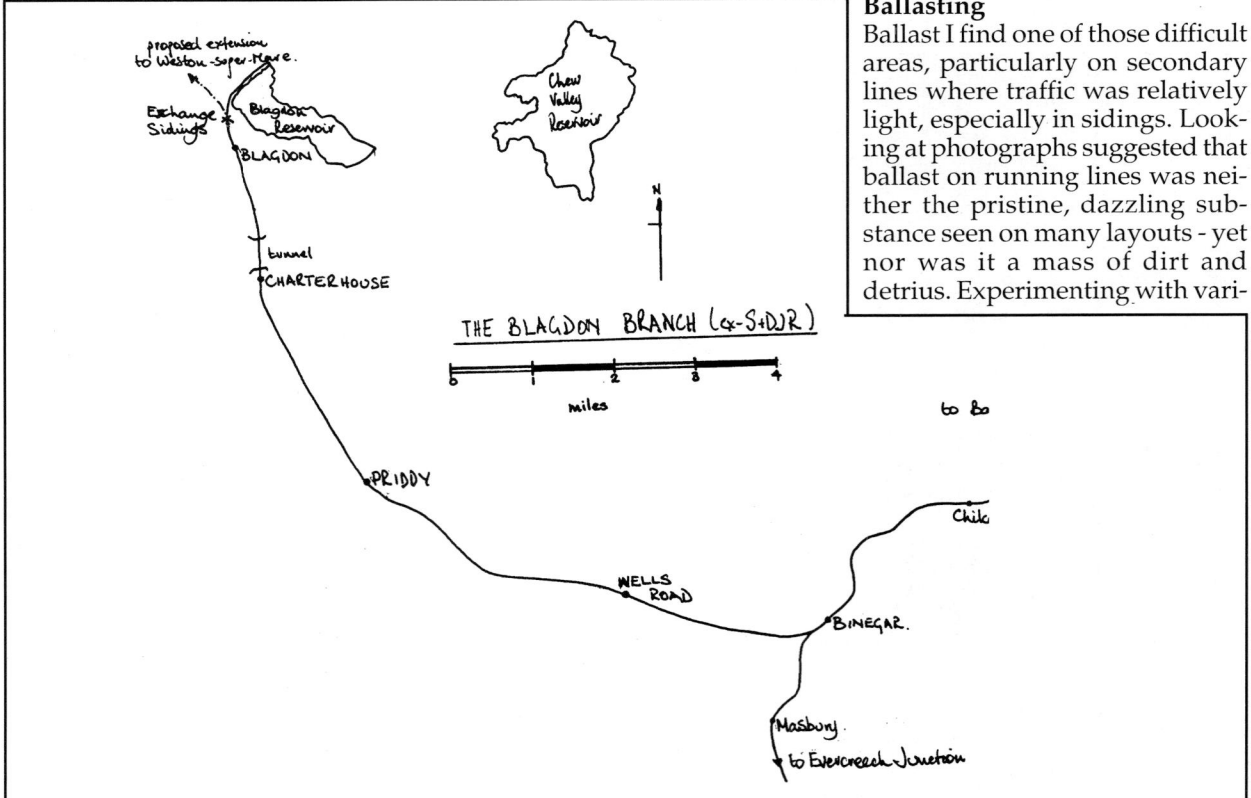

Ballast I find one of those difficult areas, particularly on secondary lines where traffic was relatively light, especially in sidings. Looking at photographs suggested that ballast on running lines was neither the pristine, dazzling substance seen on many layouts - yet nor was it a mass of dirt and detrius. Experimenting with various materials led to a mix of fine stone ballast (that sold ostensibly for 2mm. scale) and used and dried coffee grounds (Sainsbury's Viennese with fig seasoning, medium ground for cafetières!) that looked about right to my eye; in places where engines might stand and spill cinders, oil and the like, I increased the ratio of coffee-grounds to stone.

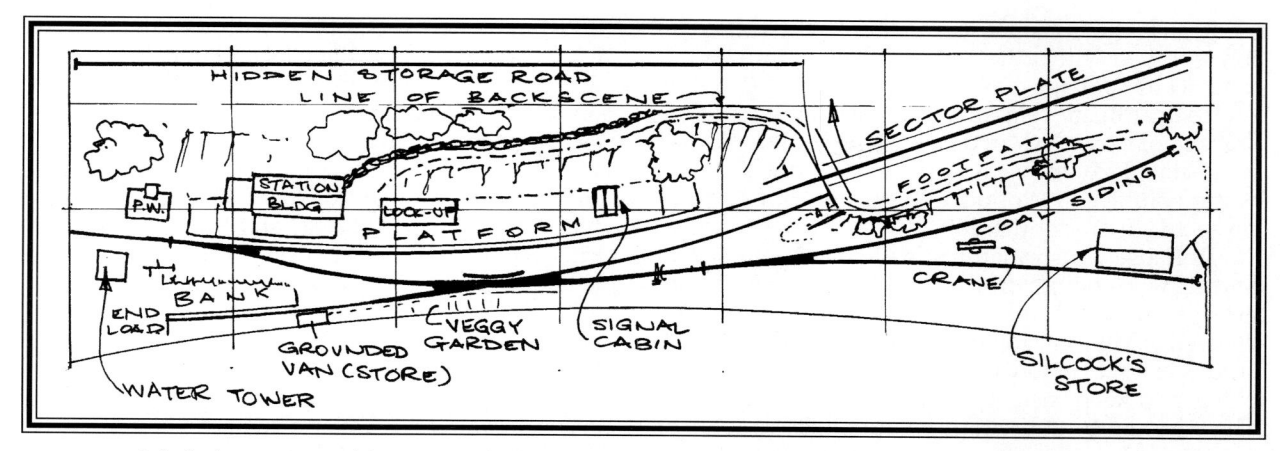

Sidings, insofar as they were ballasted at all, had an increasing mixture of coffee grounds mixed with Woodland Scenics earth and ash granules. The coal road was covered with fine coal particles sieved through a tea strainer, while the other two sidings were 'overgrown' with Heki nylon stranded grass, Woodland Scenics long grass and chopped sisal string in 5 - 10mm. lengths, all fixed with Spray-Mount and Copydex adhesives.

Electrics

Wiring of any almost sort gives me the vapours, so the aim was to keep everything simple. Experience years ago with Codar controllers (remember them?) gave me ideas of inertia control, but it seemed on reflection that this facility would be under-used on so small a layout. So I settled for my old AMR hand-held unit, with power from one of the same firm's transformers mounted within the frame of one baseboard (an instance where the deep structure of a Norman-type board comes in handy!) and suitably boxed-in.

Everything else electrical is dead simple. The controls and section-switches are mounted on the front fascia of the boards, and are kept to a minimum. The only other electrical excursion I tried was in trying to rig up a Gaugemaster 'track cleaner' (*It isn't really a cleaner - more an ionising device to improve current flow in conditions of poor contact - Eds.*) I found that, with this device installed, locos would 'creep' even with the controller at zero - even RTR models without fancy can motors. I suspect that the problem may have lain with the fact that both the cleaner and the AMR were running off the same transformer tapping. In any event, the cleaner was dispensed with! Well, I did say electrics weren't my strong point...

Landforms and Lifeforms

I have tried to convey the impression that the line had to be cut into an existing landscape, so there is a fairly consistent fall in ground level from the back of the layout to the front, as well as the downward grade of the track from the reservoir end towards Binegar. Years ago I read an article (In search of 'Z' by Tony Gillam, 'Model Railways', November 1975) about the importance of the vertical element in model landscape; this influenced my thinking quite a lot, and I determined to avoid any suggestion of the 'table top' flatness that bedevils so many model railways.

The basis of my landscape is offcuts of 50mm. thick expanded-polystyrene insulation (another skip!) glued together with Copydex and trimmed to shape with the breadknife. This was covered in an earth-mix consisting of 2:1 Polyfilla with coffee grounds (we drink a lot of coffee in our house!). The grounds add texture to the mix, but add too many and it becomes weak and crumbly. Once this earth-mix had dried, it was painted with Rowney 'Cryla' artists acrylic colours, and grassed areas were treated with Heki electrostatic nylon grass held in place with spray adhesive.

Blagdon's goods facilities are fairly sparse, consisting of the end-load dock opposite the station and these two sidings at the Binegar end. The far road is used by local coal merchants, while the nearer road is for general traffic. A Ratio crane is provided, and Silcocks have an asbestos-sheet-and-timber store for animal feeds - this was built from an LMS Society drawing. The footpath to the village crosses the running line by footbridge and runs along the bank at the rear of the yard.

Bushes are horsehair (from an old BR seat-cushion) sprayed with adhesive and dunked into Woodland Scenics foliage mat suitably torn up so that only the foam foliage remains. The complete items are stuck in place with Copydex. Making bushes like this is spectacularly messy and the spray adhesive isn't water-soluble, so I found it best to work in a well-protected area (or outside!).

Trees use wire armatures made from 7 x 7 strand fencing wire - bought from an old-fashioned ironmongers - soldered together, with the trunks and branches covered in Milliput. Foliage is, once again, from Woodland Scenics. I try to work from photos for tree-modelling, and keep basic dimensions like height correct. The wire can be cut with miniature bolt-croppers or hardened-jaw side cutters (ordinary modelling side-cutters aren't man enough). The wire is annealed with a gas flame to soften it for branch formation - don't overdo this, or it becomes very brittle.

Buildings and Details

One of the problems of modelling a fictitious location is selecting suitable structures since these are, after locos and stock, the main clues to the railway's ownership. Since there were no branches off of the S & D Bath Extension, I didn't have much to go on but, falling back on my carefully-mapped history, I reasoned that if Blagdon had been built by the 'Joint' in 1885 then the structures would probably have resembled those built only a decade earlier on the main extension. (Given that this predates the Light Railways Act of 1896 - had the branch been built 'light' then the buildings might well have looked very much like the corrugated iron structures of the real Wrington Vale Light).

A station as small as this needs to be carefully planned so as not to overwhelm it with too many structures, or structures that are too large. I did consider a large mill building as part of the goods yard, but eventually went to the other extreme with a small wood-and-asbestos feed shed based on a drawing in the October 1985 Model Railways. Similarly, I followed a small 'typical' example for the station building and other platform structures - details of their various S & D origins are in the picture-captions. Plastic was used for structure-modelling - both Evergreen sheet and strip and the Wills Scenic Series moulded panels. Ratio's goods-yard crane was acceptably close to that at Midford, while the grounded van is a Slaters MR 8-tonner, not too unlikely.

Detailing was also applied with restraint, to avoid an 'overcrowded' look: Ratio telegraph poles and fencing, Smiths S & D cast-iron notices and speed restriction signs, a Langley K6 phone box, Mikes Models water crane and Tiny Signs enamel advertisements. The signals used MSE components - the rail-built starter is from Burnham-on-Sea. There are few figures - Blagdon isn't a busy place - and I couldn't resist the Original Omnibus Bath Services Bristol/ECW L5G single-decker, even if it is way off course here!

Locos and Stock

The layout is supplied, like the real S & D, from a stocklist rather greater than it would itself justify. As part of a grander S & D Bath Extension layout project, I've been collecting/building suitable locos and stock for a good number of years, so I can produce just about anything likely to have been seen in the neighbourhood. Likely locos include the ubiquitous Ivatt 2MT tanks (choice of 2), Johnson 1P 0-4-4T, BR 3MT 2-6-2T (Kemilway kit), a Dapol 2P reworked as Bath's 40568, 'Armstrong' and 'Bulldog' goods (4F and 3F 0-6-0's to the non-S & D world) or a Bagnall tank - alias a Jinty. Less likely are the 7F 2-8-0 and LMS class 5...

The regular branch passenger set consists of a pair of non-corridor coaches (compo + bk. 3rd.), usually of ex-LMS origin. Other coaching stock, LMS constituent or ex-LSWR, makes odd appearances. Goods stock is a typical 1950's-era BR hotchpotch of wagons ranging from pre-group and ex-PO survivors to new all-steel BR standard vehicles. Exotica like special perishables vans, weltrols, or special-traffic wagons of most kinds are eschewed as untypical. Even so, common need not mean dull, and I find constructing this type of everyday stock as individual models with character and variety most satisfying.

Conclusions

A layout like Blagdon is probably most satisfying as an exercise in creating something *credible*, which can then be operated - if not to an actual timetable, then certainly to a meaningful sequence reflecting the character of the prototype. Though this is rather a personal matter, I find it most rewarding at those times when, looking at the model thus produced, I can suspend disbelief and become convinced that the S & D really *did* get to Blagdon; if I can convince a few others, then so much the better...

The 11.39 goods arrives from Binegar, trailing the Blagdon branch goods brake M328337, an LMS Dia. 1657 vehicle with extra diagonal bracing (source: LMS wagons Vol. 1) and duly branded. This wasn't common on the LM region, usually less possessive about brake vans. The guard is one Jack Dando - a descendant of Foxcote's famous signalman of the 1870's, Alfred Dando.

WIRING *For the Electrically Illiterate*
Control Panels, Wiring Techniques

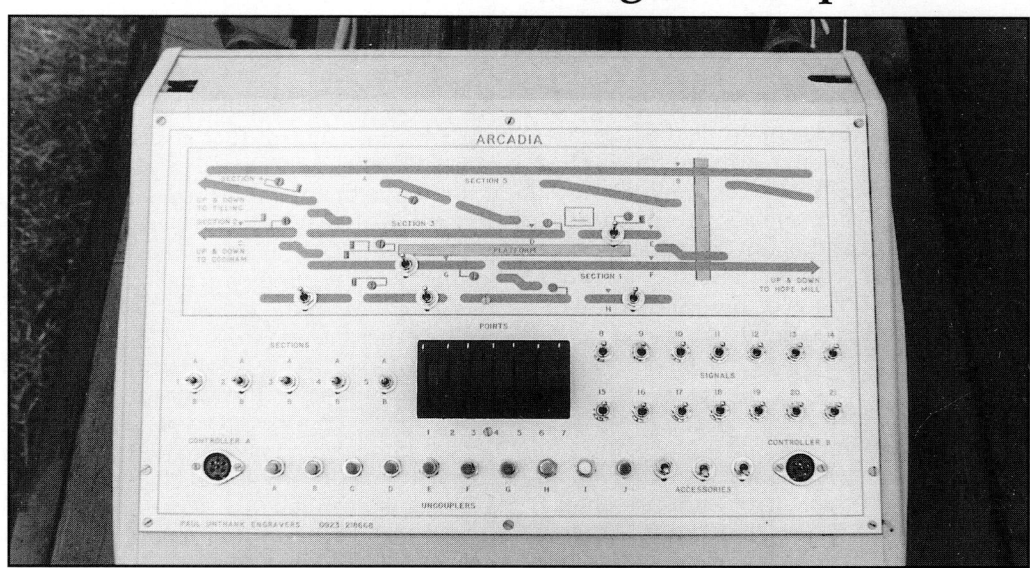

'Arcadia's' control panel is of 'signal box diagram' design, has an engraved panel from Paul Unthank Engravers.

*In this fourth section of his treatise on layout wiring, **Martin Brent** looks at the design and construction of control panels and their associated wiring, and suggests a code of practice for reliable layout electrics.*

Control Panels
As with most aspects of model railway electrics, the basic rule is: keep it simple. In my view, the answer is to follow the prototype as far as possible, using signal box practice adapted to the model requirements. There are those who will follow prototype practice slavishly and have everything, including traction power, interlocked with the signalling by using the extra switches on the point and signal motors.

To a degree, this is a commendable practice - but remember that when the unusual shunt was required in real life and there wasn't a signal to control the movement, the signalman issued verbal instructions and used flags to show when it was safe to move. If you do go down this prototypical road, unless you provide a signal for every possible eventuality rather like the old North Eastern Railway did, you'll have to have a traction power feed override switch or some other arrangement to cater for that one-off movement. My advice is to keep things simple in the first instance and add any complications of this sort later, after you have proved the basic wiring.

This is not to say that it is not possible to have a prototypical 'signalbox' control panel. A box diagram shows the stretch of track controlled by that box - all the points and signals, fouling bars and facing point locks. I won't go into detail on these terms as others are far better qualified than me to write about signalling. Suffice it to say that a lot of them are things we don't have on our models - I mention them simply to illustrate the information provided on the prototype diagram. However, the box diagram also shows salient features we *do* have, such as platforms, level crossings, etc. These aspects of the diagram can be reproduced in miniature, adapted to show other features important to us but which the prototype does without, such as the positions of isolating sections and uncouplers.

In a real signal box, the diagram is usually presented as an overhead display above the levers. This is usually not practical in model form and here, perhaps, thought should be given to the ergonomics of the control panel, just as was done on the prototype when power boxes were introduced. All too often, model railway control panels are bits of hardboard pinned vertically to the front or back of the layout. This is not good practice unless the layout is at eye level - look at any professionally-designed control panel in industry and it will almost certainly be a display slightly angled from the horizontal.

Fortunately, ready made boxes of this pattern are available new from Maplins or Radiospares or, again, try Proops for something 'pre-used'. I did, and for £5.00 obtained an old switchboard console which was ideal for the purpose. The insides were stripped out and for 'Arcadia', my new 7mm. layout, the box was turned upside down so that it presented a nice angled face to the operator. I was lucky enough to find a substantial steel strip bolted inside the case on the face that would be against the baseboard side member. Holes were drilled through the case and the strip was drilled and tapped $^{1}/_{8}$-in. Whitworth to fit some lovely large wing nuts that I acquired somewhere or the other. Nowadays, one would use a metric equivalent and fit, say, M6.0 bolts readily available from your friendly ironmonger.

Control Panel Construction and Layout
The base of my second-hand console, now the top, was cut out using a jig saw, and a professionally engraved panel in plastic laminate material commissioned from Paul Unthank (Tel: 01923 218668). This took the form of a signalbox-type diagram with all points marked in their 'normal' position (i.e.., as they would be with the levers 'back in the frame') and

with the signals shown. This was well worth the £30.00 it cost, as it provides a clear, durable panel which is wipe-clean. All of the point and signal switches were numbered so that, wherever possible, a route could be set by pulling off a numerical sequence of levers or switches.

Section breaks were also shown, the sections being identified by letter rather than number, as were isolating sections. The latter, by the way, were the only switches to be mounted actually *on* the diagram; all the others were mounted in ready cut holes (Paul has a machine for doing this) *below* the graphics. Uncouplers are the other pieces of equipment marked on the diagram and again I used letters for identification - prefixed with the letter U - to avoid confusion with either sections or the point and signal switches. The resulting panel can be seen in the picture at the head of this article.

Other Design Considerations

As already remarked, there is often a temptation to make control panels, especially the panel display, over-elaborate simply because it looks impressive. I have found, though, that it pays to make sure that it is not too cluttered and that there is room for legible labelling. With 'Arcadia', I made the mistake of not leaving quite enough room for this, making things a touch cramped.

Another design error was to place the sockets for the plug-in hand-held controllers in the panel - the sides of the casing would have been much better as the controller wires and plugs, which are apt to obstruct the panel switches, would then have been out of the way, and there would have been more space on the panel. For connecting the hand-held controllers to the panel, I used DIN audio plugs which come in a variety of sizes and pin arrangements. It seems that most people use 180º five-pin DIN's - which are readily available from any of the suppliers previously mentioned. There is a wiring convention for these plus suggested, I believe, by the EMGS, which you may care to follow while some manufacturers, most notably Gaugemaster, use either 4-pin or 5-pin 270º plugs. My own wiring convention was fixed after discussion at the local Club, so that we could all borrow each other's controllers - something worth bearing in mind for exhibitions!

The internal wiring of Arcadia's control panel incorporates my current 'best practice'. Note particularly the generous wiring runs, with wires kept slack but neatly bundled. The crossing-over of different wires and circuits on the underside of the panel itself (top picture) has been kept to a minimum, and all the tags on the switches are easily accessible to a soldering iron should switch replacement ever be necessary.

The lower picture shows the use of tag-strips and labelling to identify circuits and make fault-finding easy. The box in which this is all contained started life as a telephone switchboard, and came, secondhand, from Proops.

Wiring the control panel

When to wire the control panel is a bit 'chicken and egg'. Do you wire the layout a bit at a time and take each bit of wiring back to the panel and, in effect build up the panel and check it out as you go? Or do you complete the panel wiring and the layout wiring separately, and join them together as one operation?

Many people use the first of these approaches, but I would suggest this not the way to go. In my view, it is not conducive to a planned approach and if you have planned your operational requirements and associated wiring properly, there is no reason why you should not wire the panel in its entirety, connect it to the baseboard tag strips already labelled, and then wire in the various feeds to their positions on the other baseboards in a logical sequence. That is the practice I have adopted for my last three layouts and I find it by far the easiest. I like it because it is possible to check out the panel independently using the meter and, knowing that the panel is right, it is easy to make sure that each piece of wiring is correct as it is installed.

All the wiring in the panel box is colour-coded and terminates or originates from labelled tag strips. The feeds from the transformers in the power box (see part 1 of this series in *Digest 1*) come in on a separate plug and the outputs - all fifty-something of them - go out on two multi-pin plugs to the baseboard. When wiring a panel, do remember to allow for the panel top to be raised for maintenance. Include at least six inches 'slack' and tidy the wires together using cable ties or tape. Fourtrack Models do a natty line in spiral wire-wrap which is ideal for this purpose, as short lengths can be easily slipped over the wiring bundles and other wires added or

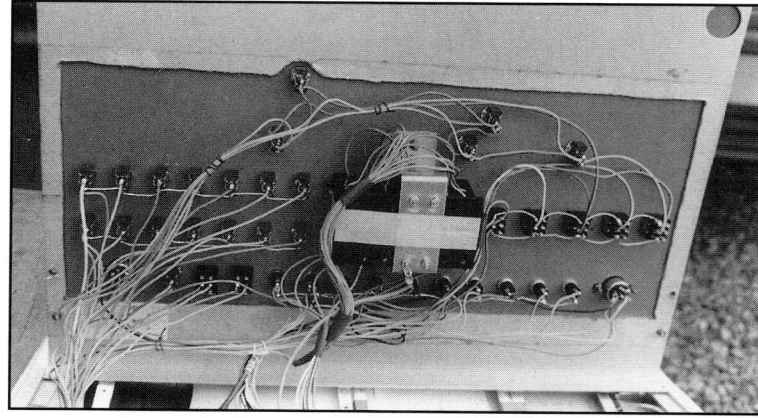

Brent's Rules of Wiring

Before dealing with the nitty-gritty of wiring practice, a few words on how we can ensure that the wiring is reliable and worthy of the model might be in order. It my basic credo that wiring should be fitted and forgotten. Certain criteria are needed if this is to be the case.

The easiest way of ensuring this happens is to promulgate (corr, this lexicon thingy on my computer really works!) a set of simple rules. Consistency of supply is all-important on model railways - most problems of reliable slow running can be traced to dirty track, dirty wheels, poor pick up or inconsistent supply. I could wax lyrical (*Surely not lyrical, Martin; more like wearily, given the topics... Ed.*) about the first three of these evils, but will here confine myself to the last-mentioned. There are two things to avoid; excessive voltage drop, which many people worry about far too much, and its first cousin, high resistance - especially in joints -which most people don't worry about nearly enough!

The former, as discussed in Part 1 of this series, can be dealt with by using a sufficiently heavy gauge of wire, consistent with the loads being carried. Avoiding the latter is a case of a taking a little bit of extra care and effort. High resistance means exactly that - our poor little 12 volts (or less) can come up against a poorly-soldered joint (usually what is called a 'cold' or 'dry' joint), or encounter that temporary connection that you made by twisting two wires together and have never got round to doing properly. The result is that it will lose most of its 'urge' in forcing its way past this electrical blockage. The result is less consistent voltage and less overall power at the motor where it is needed; too much of this, and your locomotive will hardly be able to pull itself along.

High resistance can occur at a number of points in model railway wiring - particularly in the actual track, still part of the circuit, remember. Most beginners rely on commercially-available fishplates or rail joiners to connect their rails together and to give a continuous supply of juice. Such things may be O.K as locating devices, but they are not so hot as electrical connectors, being often a prime cause of high resistance joints. We have already discussed cab control and common return wiring, but even when using this system it is essential that proper connecting techniques are used - which means each bit of rail gets a separate connection.

This may mean a lot more wiring, but reliability will be as near 100% as you can get it. The easiest way of achieving this reliability is to use a heavy gauge supply wire under the baseboard, to which the various lengths of rail are individually connected. In industry, this supply wire would be called a busbar. I use old household single core earth wire for this purpose, other modellers use old rail, or wire each piece of rail back to a common tag strip.

Taking the supply to the rail is simple. If you are using commercial flexible track or PCB-sleepered points, the easiest way is to make some droppers, as described in part 1, to get the feed up through the baseboard. Under the board, a wire - properly colour-coded - can be run from this dropper to the relevant bus-bar, or you can take your bus-bar to the dropper - I prefer the latter but the choice is yours.

The bus-bar itself I then take to a tagstrip, which I use to locate all the wire-ends neatly and also as a distribution point for cross-board-joint circuits, all circuits within a section, or other units of wiring that need bringing to a common point. One tip that I have found useful but not seen mentioned before: paint the underside of the baseboard with white emulsion, or apply large sticky labels next to these tagstrips, busbars or W.H.Y., and you can then use a felt-tip pen to label what the wires and other bits really are.

All this can be summarised under five basic headings, as a set of rules:

Rule 1: Every discrete piece of rail should have its own supply connection - no piece of track should rely solely on fishplates for electrical continuity.
Rule 2: All wires should terminate on a labelled tagstrip and each wire should be clearly identified at each and every junction.
Rule 3: All wires should be colour coded and the colour code should remain constant from the control panel to the equipment.
Rule 4: All joints should be soldered or screwed using proper screw connectors. All joints should be tested, not only electrically but also mechanically immediately after installation- by giving them a good tug. If soldering, ordinary multi-core solder is the thing. Do *not* use acid flux on wiring.
Rule 5: Make a wiring diagram and *keep it!* Any amendments should be recorded thereon.

Next Time: Designing control circuits.

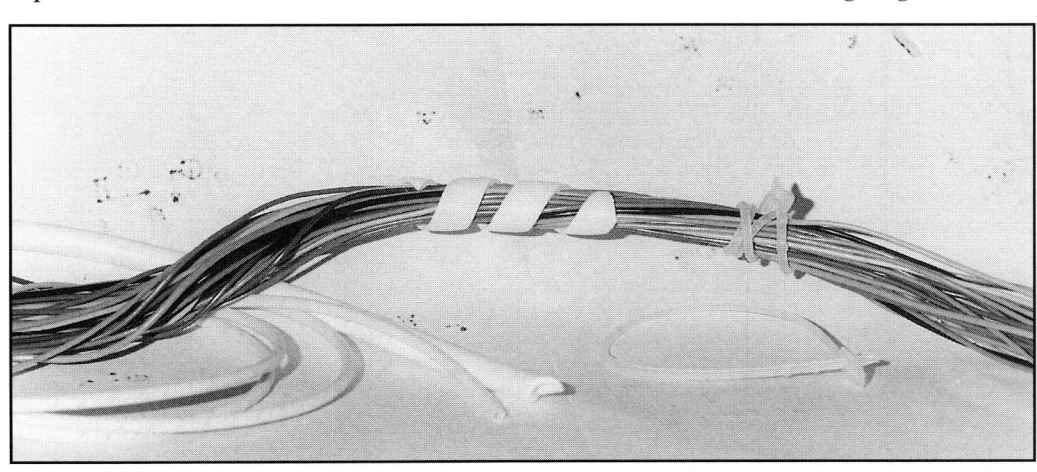

Wiring aids: Coil trunking is available in variety of sizes, as are cable ties, both great aids. Both these items are available from Fourtrack Models.

Carriage & Wagon Department

The Wagons of the Denaby Collieries

A new recruit to the team that produces our regular rolling stock feature, **Chris Crofts** *is a skilful modeller of wagons as well as being a well-respected researcher.*

A wagon from the series built by The Derbyshire Carriage & Wagon Co. in 1936. The tare weight of this particular example was 7t 5c
Photograph C.N. Crofts Collection

The wagons of Denaby and Cadeby Main Collieries Ltd. (after 1936 Amalgamated Denaby Collieries Ltd.) must have been a familiar sight in the 1930s. The livery was striking, and there were lots of them, at least 1,350 being built around 1912 – 13 and some 1,800 from 1935 onwards, in addition to earlier batches and wagons purchased second-hand. The earlier, 10-ton wagons retained the lettering *Denaby Main Colliery* across the top of the wagon, at least up to 1937. It is, however, the 1930s wagons that we are concerned with here.

Denaby wagons have featured before in the model railway press; there were sketches in the Model Railway Constructor in 1936 and the Model Railway News in 1946, neither suitable for building models to the standards required in 1996. A post-war photograph of a Charles Roberts example appeared in the Railway Modeller for April 1963, and a short item by Peter Matthews, with a works photograph of a Derbyshire Carriage & Wagon Co. wagon, in the Model Railway News for July 1964. The last-named source was by far the most useful, but no dimensions were given, so that when, in 1974, I attempted a model, I had to try to scale the height of the wagon side from the depth of the headstock, resulting in a wagon a scale $3^{1}/_{2}$in. too high. Twenty-two years later I could build an accurate model from just such a photograph, but for the benefit of the newcomers of today I shall give dimensions. Peter Matthews also gave No. 2498 as identical, and I built a model of it, but it was almost certainly not by the same builder and so would have exhibited subtle differences which I would now wish to reproduce.

And so to the wagons, beginning with the Derbyshire Carriage & Wagon Co. batch. These, and all the others detailed in this article, were built to the 1923 RCH specification and were known as Standard wagons. Dimensions were :-

Length	16ft. 6in.
Width over body	8ft. 0in.
Width over headstocks	8ft. 1in.
Depth of body	4ft. 7in.
Width over solebars	7ft. 1in.
Solebars and headstocks	12in. x 5in.
Depth of siderail	$5^{1}/_{2}$in.

Thirty wagons were inspected and registered by the L&NER (Nos. 10155 and upwards) on 17th November 1936 and a further five (10200 and up) on 22nd December. (Just four days before I appeared on the scene!) Details are as follows :-

No.	Tare	No.	Tare	No.	Tare
2600	7-5-0	2610	7-5-0	2620	7-5-0
2601	7-5-1	2611	7-5-1	2621	7-5-2
2602	7-4-2	2612	7-5-0	2622	7-5-3
2603	7-4-3	2613	7-4-1	2623	7-5-3
2604	7-4-3	2614	7-5-0	2624	7-5-1
2605	7-5-0	2615	7-5-2	2625	7-5-2
2606	7-5-0	2616	7-5-1	2626	7-5-3
2607	7-4-2	2617	7-5-0	2627	7-6-0
2608	7-5-1	2618	7-5-1	2628	7-6-3
2609	7-4-1	2619	7-5-3	2629	7-6-2
2645	7-7-1	2647	7-6-0	2649	7-6-1
2646	7-6-1	2649	7-6-3		

Another example of the wagon, this time No. 2680 as built by Edward Eastwood, of the Railway Wagon Works, Chesterfield.

Photograph C.N. Crofts Collection

I do not know what happened to Nos. 2630-44. Perhaps I missed them when I made a necessarily quick search through the register, or they may have been registered by the LMS (registers for 1936 believed destroyed), or they may have been lettered *Maltby*, or perhaps, for some unknown reason, those numbers were not used. Whatever actually happened, we have here 35 wagons which were as near identical (except for the tare weights) as any modeller will be able to make them.

The wagons were red; LMS crimson followed by matt varnish should give a reasonable effect, or some bauxite could be mixed in. The final colour, though, should be one that the average person (approximately 92 - 94% of males apparently have perfect colour perception!) would call red rather than brown. The letters, white shaded black, were applied by Cecil Kipling. Although stencils were not used, such was his skill that all the wagons can be taken as identical for modelling purposes. The small letters (tare, load, empty to ....) were unshaded, the Cc black on a yellow background, and the star yellow – not white, as one often sees on proprietary models. Ironwork was picked out in black, as were the ends of the headstocks.

A few constructional details are worthy of note. The end door had an angle iron protector plate, bolted to the bottom plank, hence the row of nuts and washers on that plank. The end door bar went through the top plank of the sheeting on both sides, and was secured with 2in. Whitworth nuts – about $3^1/_8$ across the flats – big enough to model in 4mm. scale. The side knee washer plates were not bent round. The strap-bolts had separate washers. Brackets were fitted between the side rail and solebar. Nuts were hexagonal.

The next wagon, No. 2680, was built by Edward Eastwood, whose works were on the down side of the Midland main line just north of Chesterfield station, and about 2 miles south of the Derbyshire Carriage & Wagon Co.'s works. It was probably registered by the LMS and recorded in one of the lost registers, so it is impossible to give full particulars of the batch. Although the wagon is basically similar to No. 2624, there are numerous small differences in construction and livery; it is these differences that make the study and modelling of wagons so fascinating.

The end door has no angle iron door protector and hence no row of nuts. The strap-bolts have a washer plate rather than individual washers. Side door thresholds are present–the half-round pieces of wood to close the gap between door and side rail when the door is down. Just below the D of Denaby, one can see, on the original print, that the end door fastener bolt is fixed to the sheeting by two bolts, in addition to the one through the end knee. Most builders used only one, although the RCH drawing showed two. Nuts are square, rather than hexagonal.

There are slight differences in the livery. The number, 2680, is rendered in rather taller figures than 2624, and the Cc and star are placed differently. Readers will doubtless note other minor differences.

We now turn to the wagon built by Charles Roberts of Wakefield, an altogether bigger concern than those previously considered. They turned out wagons by the hundred rather than by the ten.

A photograph of No. 3189 appears in Bill Hudson's *Private Owner Wagons, Volume Four*, and, as the book is still in print, there is no point in using the same photograph here. Instead, I have chosen No. 3111, which has a virtually identical body to No. 3189 but on a steel frame. I strongly suspect that this was one of the batch of 50, numbers unknown, referred to by Bill, and as they immediately preceded Nos. 3150 - 3239, I would guess that their numbers were 3100 - 3149.

Apart from the steel frame, there are small differences in construction between this wagon and those already noted. The end door has the angle iron protector and row of bolts. There is only one bolt holding the end door fastener bolt (in addition to the one through the end knee). Side knee washer plates are bent round, rather more sharply than on a wood framed wagon because of the smaller depth of the side rail. Charles Roberts always screwed the side door protector to the door, rather than bolting it on. Countersunk screws were used, so that in a model the protector is left plain. The bottom door casting or bracket is different on Charles Roberts wagons, having, apparently, a half - round front instead of the "open" pattern used by Derbyshire and Eastwoods. The buffers at the door end have (in wood - framed wagons) a straight top rib. Strap-bolt washer plates are used, and nuts are square, except, oddly, for a few on the solebars of wood-framed wagons.

Issue No. 4

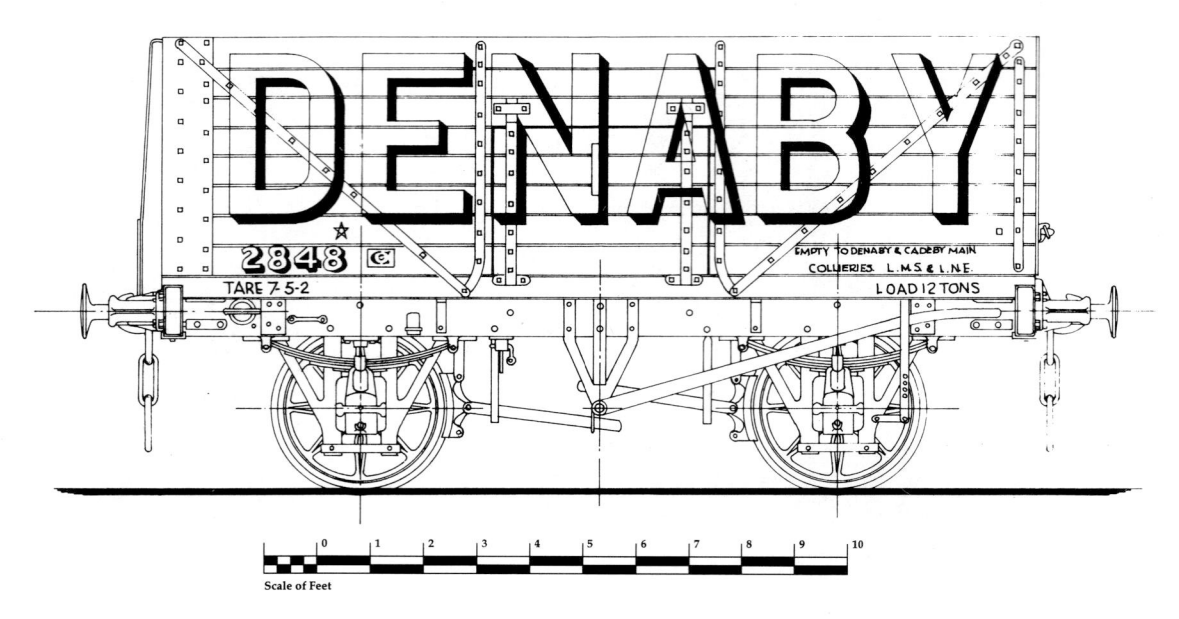

Turning to livery, it will be immediately apparent that the D and the B are much more rounded than those on the Derbyshire and Eastwoods wagons. On No. 3111 (but not on 3189) the top of 'Empty to' is on the second plank up. Again, the Cc and star are differently placed; this placing seems to have been consistent within a batch but could differ between batches. A peculiarity of Charles Roberts wagons was that the whole of the door bang spring was painted the colour of the solebar, black for steel frames and (in this case) red for wood frames.

For my drawing I have chosen No. 2848, but seen from the opposite side from the photographs of 3111 and 3189, i.e. with the end door to the right. This shows how the letters had been spaced differently on the two sides. However, don't just take my word for it; the wagon appears in Charles Roberts photograph No. 2819 which should be available from the National Railway Museum or, rather more cheaply (but from a copy negative) from the HMRS.

Details of identical wagons that can be made out on the photograph are: -

No.	Tare	No.	Tare
2818	7-6-0	2843	7-4-?
2817	7-7-0	2842	7-?-?
2844	7-6-2		

This gives a total of 36 wagons, including No. 3189 in Bill's book, with fairly full details, including tare weights. This should be enough for most people, but as a final offering I show one of these wagons photographed by Gordon Coltas near the end of its life, in 1956 or 1957. The question is, though, who built it?

The end door bolt is fixed by two bolts, so that rules out Derbyshire or Charles Roberts. There is an end door bar protector - a small strip of iron to stop the bar working out if the nut on the far side should be lost; there is no door threshold (or bolt holes to show where it might have been, and the wheels are spoked, not disc. These three together

A Charles Roberts example, numbered 3111. The Cc and star markings are clearly seen at the left-hand end of the wagon.

Photograph Chas. Roberts/C.N. Crofts Collection

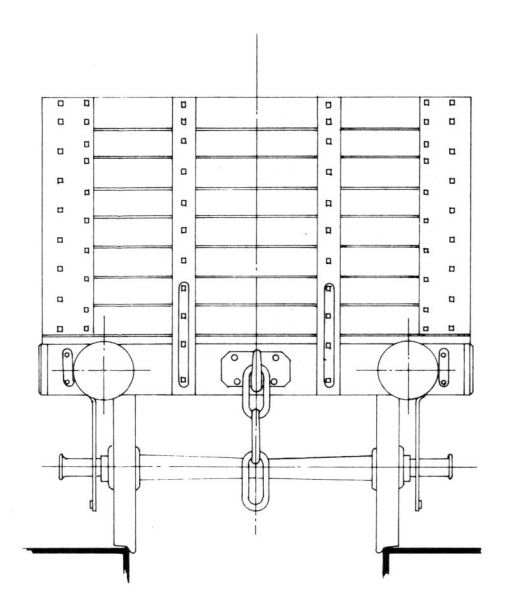

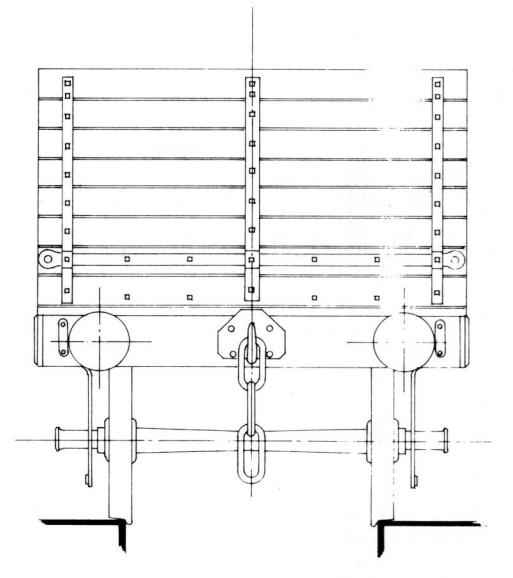

Denaby wagon by Charles Roberts. Drawing by Chris Crofts & Mike Peascod.

would seem to rule out Eastwoods. The lettering (what remains of it) has the rounded D and B rather like the Charles Roberts wagons, but the middle stroke of the E is wholly on the fourth board down, unlike the wagons of the builders we have considered where the middle stroke extends upwards onto the third board. All these features indicate that another, unknown builder was involved, but I haven't enough information to say which one.

Running Period
Obviously the wagons detailed above would only have been seen from 1935 - 6 onwards, but older wagons would have been repainted into this livery at an earlier date. The earliest photographic evidence that I have of the use of the large Denaby is dated May 1932. However, Charles Roberts built a mixed batch of Denaby and Maltby wagons in 1930; the Maltby wagons received the large lettering and it seems likely that the Denaby examples were similarly lettered. I have a photograph of a Charles Roberts Maltby wagon dated 1927 with the large letters, and I feel that Denaby wagons would be receiving them at this period, but I wouldn't like to go much earlier, certainly not to pre-grouping days.

Shall we see models built as a result of this article? I hope so, after all, I built two as a result of Peter Matthews' article. Nowadays you have the Slater's kit to help you (but only in 7mm. scale); with small modifications it would make up into any of the wagons I have detailed, except, of course, the steel framed ones.

Standard Markings on PO Wagons.
Cc and **yellow star** on the wagons both indicate that the owner of the wagon participated in a commuted charge scheme. In the 1926 scheme (Cc) the owner paid a small annual charge in lieu of siding rent. In the 1933 scheme (star) the annual payment covered empty haulage back to base, or repair shops (for green carded wagons).

A time-worn Denaby wagon, with B.R. standard markings painted on black patches over the original Private Owner livery. Photographed at Carnforth M.P.D. in 1956 or 1957.

Photograph Gordon Coltas

Issue No. 4

69

M.P.D.

Project 2251 - Finishing Off

As you will all probably have gathered by now, our 2251 project has been brought to you 'live'; that is, we have been working on the model, writing about it and taking photographs very much as we went along. Which is how we came to tell you, at the end of the last instalment, to fix the fall plate to the loco footplate - it was what we'd just done. But, on taking up the cudgels to finish the model off for this final article, it was very quickly realised that this was a no-no; with the fall-plate fixed to the loco, and with the loaded tender needing to have it's hook lifted over the loco drawbar for coupling-up, something had to give... The fall-plate is now fixed to the *tender*, and no further trouble has ensued. Phew!

Fabricating Bufferbeam Hoses

This was the next job tackled. It always amazes me (this part of the job being down to Rice) just how many people seem to have a blind spot over vacuum and steam-heat hoses - the same old commercial offerings turn up on loco after loco, whether they're suited or no. Considering how easy it is to make accurate, convincing and properly-configured hoses, I'm at a loss to understand this.

So - the ingredients: simple - first, some suitable wire for the 'core' - I usually use Eileen's 0.9mm. half-hard brass (comes in a coil) although you can use the straight-drawn stuff as purveyed by Alan Gibson. For the 'windings', I use the 0.2mm. copper wire that is simply obtained by stripping down about a 4-in. length of some ordinary layout hook-up wire (7/02 flex, as recommended by Mr. Brent).

Method: Take a good length of core wire, and bend the shape of the vacuum connecting hose itself into one end, working from a photo. Don't, at this stage, form any sharp angles, as at the top of a vacuum standard. Now take a bit of the 0.2mm. copper wire, and wrap it round the core wire at the opposite end to your bend, Now continue wrapping it tightly round the core to form a tight coil. Don't worry if you coils are spread out a bit; it's easy enough to push them up together.

Carry on coiling until you've got a long enough length of coil to form the actual vacuum connection. Make a 'double turn' at the top of this, to simulate the joint where it joins the vacuum standard. Now push the coil down the core wire onto the pre-shaped 'hose' portion, and secure with a trace of solder both ends. A short length of wire at the lower end simulates the connecting lug of the hose coupling - see the final picture opposite.

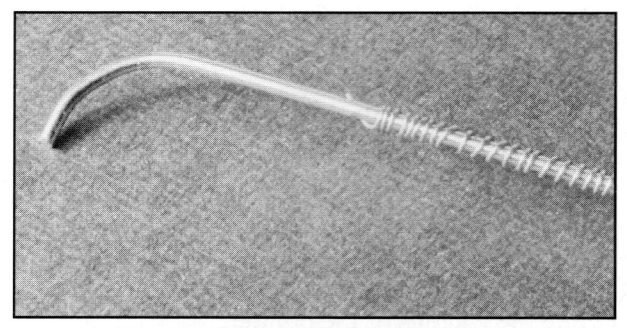

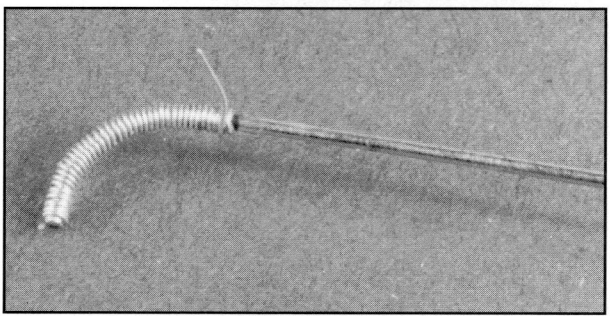

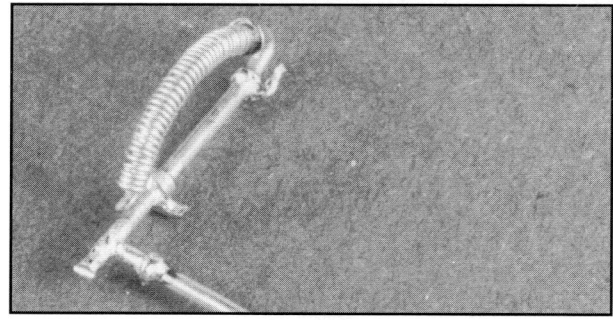

Above - the stages in fabricating a vacuum standard: Top: The core wire formed, and the hose coiled loosely on the far end. Centre: The coil closed up, moved to hose location and secured, and Bottom: The rest of the standard is formed and any other detail added. (This example is an LMS tender type, in case you're wondering where it is on the 2251!)

Last job is to complete the bending-up of the vacuum standard, and to add any further detail - fixing brackets, pipe unions, shut-off cocks and so on - that the prototype may have. On the 2251, as I often do, the vacuum fittings are attached to the chassis, being simply soldered to the end frame spacer on both loco and tender chassis.

On the 2251, we elected to model the engine in 'summer trim' - which, on the GWR, meant with steam-heat hoses removed. All that was left was the connecting flange on the end of a short piece of pipe protruding from beneath the bufferbeam. This was simulated with scraps of brass soldered firmly to the end of 0.9mm. wire and filed to the prototype's oval shape.

Melt-in Metal Details

We decided to leave the 2251's main boiler handrails alone, as Mainline don't do a bad job on them at all - better than a lot of 'scale' handrails of not so many years since. We did, however, elect to replace the moulded cab and tender grabs with wire, as well as adding the short grabs above the front steps on the loco and providing some brass-strip lamp-irons at this front end. All these metal details were 'melted in' to the plastic body, using a 15W Antex soldering iron and some simple spacers filed up out of Formica scraps. The photo below should give the gist of the process. All the brass used for these bits was chemically blackened - with Philips Gun Blue - preferably before being fitted.

Melting-in detail such as grab-irons was accomplished with a 15W Antex soldering iron and a heatproof spacer filed up from a scrap of Formica. The bit used is an old one, retired from soldering, and having two narrow V-grooves filed in its tip at 90° to give location on the part being 'melted in'.

The cab stanchions were replaced in a similar manner, using 0.45mm. hard brass wire (Eileen); the same wire was used for the cabside, tender rear and front step grabs and the tender front stanchions. The lamp-irons were Eileen's 0.8mm. x 10thou, brass strip, also melted in - GW irons - apart from the bunker or tender rear ones - are very simple. We left the moulded tender lamp-irons be.

The other metal details - the new buffers, cast-brass GW whistles, loco backhead and the replacement smokebox-door handles (Gibson turnings - very fine indeed) were attached with Cyano or Bostik Clear as appropriate. The last metal detail was the new capuchon to replace the wrongly-placed plastic effort filed off back in about part 2. This was made from a short piece of Eileen's 0.5mm. square brass formed to the correct radius around a tapered scriber cut off to size and fixed to the copper cap with cyano.

Chassis Details

Two things were missing from the chassis: sandboxes and balance weights. Both were made out of plastic, the sandboxes by laminating four layers of 1mm. thick plastic - waste from a Wills Scenic Series kit-sprue - together with MEK, then carving and filing this to

Top: The front end of the model with unwanted plastic detail carved away. This was done with a Swann-Morton craft knife, using a sharp No. 2 curved blade, and finished with fine wet-and-sry 'rubbing sticks' - 320 grit stuck to eigth-square stripwood with the ends bevelled. Care on this job pays dividends! Centre and bottom: the re-detailed front end and tender rear. Wire grabs, brass strip lamp-irons, turned buffers, turned smokebox door handles and new bufferbeam hose details are the major changes. Note simple wire loop on front of loco - all that is needed for compatibility with tension-lock couplings.

shape and melting-in sandpipes of 0.45mm. brass wire. The completed sandboxes were then secured with cyano. The reason for using plastic for these? No chance of a short if they touch anything live - and they will, they will...

The balance weights were cut out of 20thou. black Plastikard with an Olfa compass cutter, a useful gadget (you can get it from Crookes of Sheffield), although the old dodge of a pair of dividers and keep twiddling will also serve to produce some discs of plastic of appropriate size from which to cut the weights. I'm not quite sure we've got ours quite right; a set of etched ones (someone's bound to do them...) is an attractive alternative. The resulting weights were fixed - hopefully in the right place - with cyano.

Left: The completed chassis ready for painting, with home-made plastic sandboxes filed from laminated plastic offcuts and fixed with cyano - note sandpipe spot-soldered to brake-hanger to reinforce location. Balance weights cut from 20thou. Plastikard using Olfa Compass Cutter (below), fixed with cyano and sanded down to a flush fit. Note also coupling loop and front vacuum pipe fixed to chassis. Altogether, a vast improvement on the original Mainline offering!

Painting and Final Details

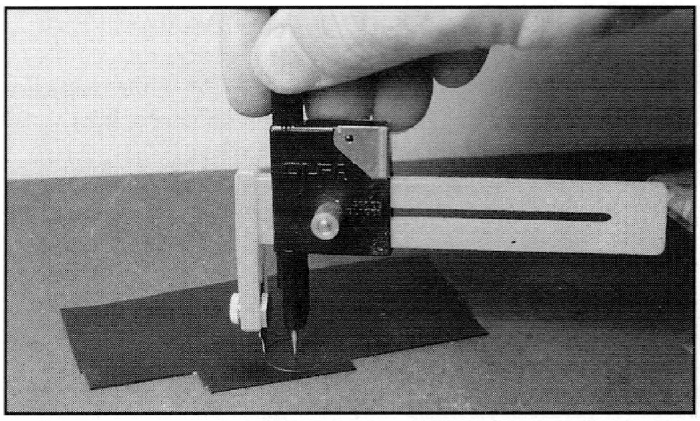

The main paint job is, of course, the chassis - both loco and tender. The loco chassis was sprayed 'under power', the motor being masked off with thin card and the chassis run slowly from a 6V battery while being sprayed with grey acrylic primer, followed by Humbrol 'Krylon' spray matt black. This was done outside on a sunny afternoon, with a cardboard carton on its side to catch the overspray. Once the paint was dry, the chassis was run up on the controller and the unwanted paint removed from the wheel treads and coupling rods with a fine three-sided scraper, some fine wet-and-dry paper and a fibreglass burnisher.

The tender chassis was also painted outside, and the wheel-treads cleaned by rotating the wheels against a rubber wheel (rubber door-stop on a bolt) chucked in an electric drill. Don't forget (as we did!) that the back of the tyres of the rear tender wheels carries the tender pick-ups and need thorough cleaning if these are to work.

On the body, only touch-up work was required. The new boiler bottom was brush-painted in Humbrol 104 GW Loco Green, a very good match for Mainline's shade. We also repainted the footplating, smokebox, cab roof and interior and tender underframe with Humbrol Hobby Acrylic, using a variety of matt black/matt earth/red oxide mixes to get 'warm' shades that weren't too strong. The bufferbeams were also repainted with acrylics, and some light dry-brush weathering applied overall. We opted for a workaday finish, also painting the top of the chimney copper cap black (as it should be) and the safety valve bonnet - duly fined-down with files - green. The new handrails were chemically blackened to match the Mainline originals. Last touches were the glazing of the cab side windows with Ferrero Rocher chocolate box and the installation of the painted cab backhead and loco crew. The result? A 2251 that looks every inch a scale model - and runs like one too!

Above and left: The finished model belies its £6 junkstall origins; it may not be the 'ultimate' 4mm. scale 2251 - although there are further improvements that could be made - but it's a good compromise for a working layout loco.

RailMODEL Digest

WORKSHOP -
The DG Coupling

*In the first of a series of articles on the whole business of auto-couplings in all scales, **John Chambers** takes a look at the DG, a device originating in 2mm. but now increasingly used across the scales up to 7mm.*

Introduction
Let's start with a riddle. What's been around for rather a long time, is reliable in operation and easy to set up, yet would appear to be undervalued and under-used by most railway modellers? The DG coupling, that's what.

I first became acquainted with this device many years ago, when operating the Luton Hoo 2mm. fine scale branch on the M.R.C.'s 'Chiltern Green' layout. Subsequently, my friend Keith Foster - not a man to waste his time and beer-money on anything less than totally reliable - started using them on his first P4 MSLR layout, 'Kenton and Aspall'. I can't remember how many exhibitions we did with this model, but I do recall what a huge contribution the DG coupler made to the entertainment value provided, its consistent operation allowing slick shunting and removing any need for 'the great hand in the sky'.

The DG Described
The DG, like the better-known Alex Jackson, is a magnetically-operated coupling with the facility for delayed uncoupling remote from the actuating electromagnet. It is not as unobtrusive as the Jackson, but it is more robust and is, I've found, much easier to fit and adjust. It can be obtained for use in 2, 3, 4 and 7mm. scales. My experience lies with the 'Type B' version for 4mm. scale, so that's what I'm describing here; the other variants are similar, differing in size and metal thickness.

The Type B comes in packs of 16, and contains coupling etches, delay latches, phosphor-bronze wire for the coupling loops and steel wire for the magnetically-actuated droppers. This is a standard quantity for all the variants, and DG can also supply suitable electromagnets and loop-bending jigs to assist in coupling manufacture. The basic couplings are etched in brass and could be chemically-blackened after assembly if required

Assembly
First job is to separate a coupling etch and a delay catch from the fret and to clean it up, removing the etching cusp from the edges and making sure to remove all the remains of the attachment tags. This coupling etch consists of the buffing-plate with hook attached, plus fold-down pivot points for the loop - see the photo of the components and the sketch for the relationship of these parts.

Construction of the coupling starts by bending the buffing-plate down through 90º, which raises the hook to the vertical; this is then bent back slightly - the instructions suggest 20º. I also like to bend the outer ends of the buffing plates back slightly, to avoid any 'catching' when wagons are being pushed. To reinforce the coupling, the instructions also suggest running small fillets of solder into the joint between coupler and buffing plate - I've never found it necessary in 4mm. scale.

The delay latch is now curved through some 30º from a point just above its triangular base and then inserted into the slot in the coupler between the pivot points of the loop; the two 'fingers' formed by the split etched into its base are then spread to keep it in place. It should rest gently on top of the coupler hook by its own weight, and when lifted should not pass the vertical - this ensures it will always fall back into the 'rest' position on top of the hook.

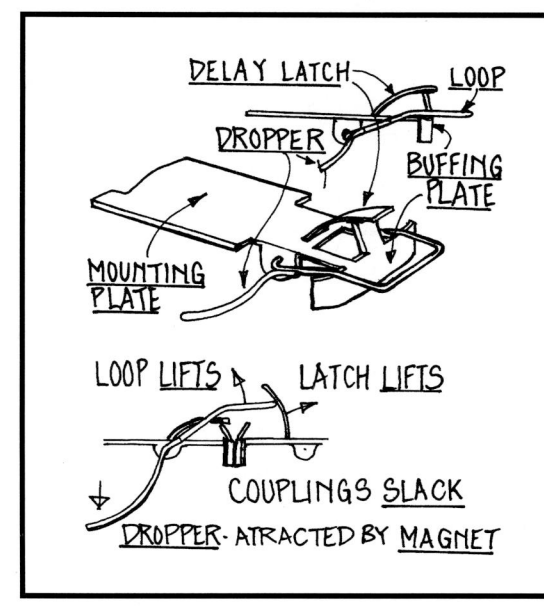

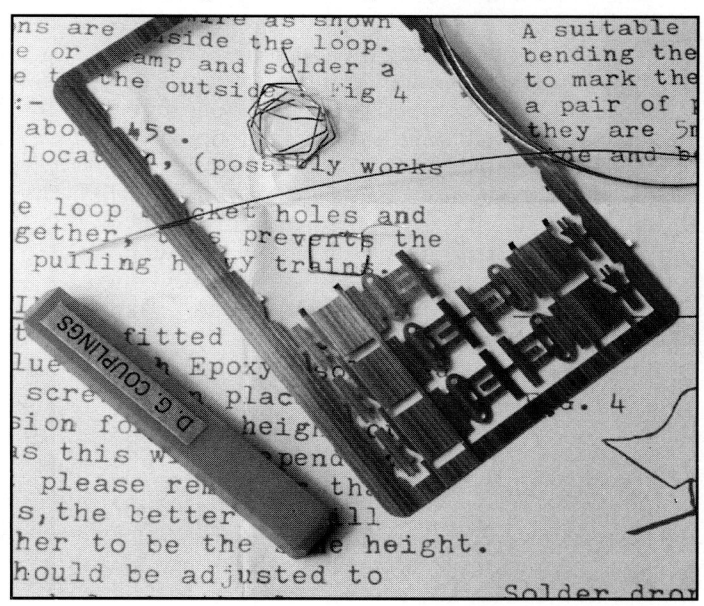

DG couplers about to engage - these are 'handed', so the mineral wagon has no loop on this end. This picture shows the compact size of the units, and the relationship of the buffing plates to the vehicle buffers - 1mm. in advance, in this case. This close-coupling distance can be varied by moving the couplings back a bit on the wagons - less than 1mm. between buffer faces is possible on easy curves.

The loop pivot ears are next bent down ready to receive the phosphor-bronze wire coupler loops and the steel actuating arm. I purchased a jig for bending these loops, a simple but ingenious device for mass-producing them very quickly. It consists of a piece of rectangular Paxolin bar of the correct dimensions - 7.5mm. x 5mm. This has a fine hole drilled in one face, and the idea is simply to insert the end of a length of the PB wire into this hole and wind it tightly round the bar, producing a sort of rectangular coil. By snipping this in the appropriate place - Xuron rail cutters are ideal - you get lots of rectangular loops.

Without the jig, it is necessary to use pliers and a bit of care. DG, in their instructions, suggest marking the tapered jaws of some fine snipe-nose pliers at the appropriate points to get accurate 5mm. and 7.5mm. dimensions. The long axis of the loop goes out over the buffing bar, so it's necessary to get the join in the middle of one of the 5mm. sides. The PB loop is 'sprung' into the pivot holes in the ears, and then the tails spot-soldered together to stop it being pulled out - DG say this is needed for heavy trains, but I've found it wise to do it anyway. The front of the loop is then bent down so that it lies horizontal where it lies over the buffing plate.

However, before fitting the loops the steel dropper wires should be added. You have a choice here, either a straight or a curved dropper. The dropper is soldered to the right hand side of the loop, looking at the coupling from the wagon end and above. For a straight dropper, a piece of the steel wire is bent through 45º and soldered to the loop with the dropper to the rear, trailing down below the wagon bufferbeam. The curved version needs a 90º bend, and the wire bending into a quarter-circle. In both cases, the aim is to get the bottom of the dropper just above rail-head level. The sketch at the right should make this clear. (They say a picture is worth a thousand words, but in this rag it don't pay as well!). The action of the dropper, which acts behind the pivot point of the loop, is to lift the latter clear of the hook. While the magnet is activated, it keeps the loop up, the basis of the delayed action. Pushing the vehicles together then turning the magnet off lets the loop drop onto the delay latch, as we shall see.

Fitting

Mounting a DG coupling on a vehicle is simplicity itself. They can be glued, screwed or soldered in place, depending on what the actual vehicle is made of - the mounting plate is quite generous in size. The instructions leave it up to the user as to the exact height the coupler is mounted at - so long as the dropper is adjusted to be just above the rails. I mount mine immediately below the bufferbeams, which usually calls for Plastikard packing to build up a suitable mounting pad. For metal vehicles, I bend up a bit of scrap brass into a flat U-shape to get a base at the same level, and solder the coupling to this.

Again, the exact relationship of the buffing plate to the vehicle buffers is left up to the individual modeller to determine. Obviously, the further out the loops are, the further apart the vehicles will couple, but the sharper the curve that it can negotiate. Common sense dictates what is appropriate for individual circumstances - and also that the mounting height and distance out from the buffers are kept as consistent as possible. A simple jig or two in Plastikard could assist in achieving this consistency. I have mounted my couplings with the buffing plate 1mm. in advance of the buffers, and my wagons negotiate a B6 turnout in either direction without their buffers getting too intimate with each other. I could probably have got away with 0.5mm., it all depends on the tightness of your curves (on the track!!)

Although consistency is desirable, you can get away with minor differences, especially in height, and the DG is both easy to adjust in this respect (just

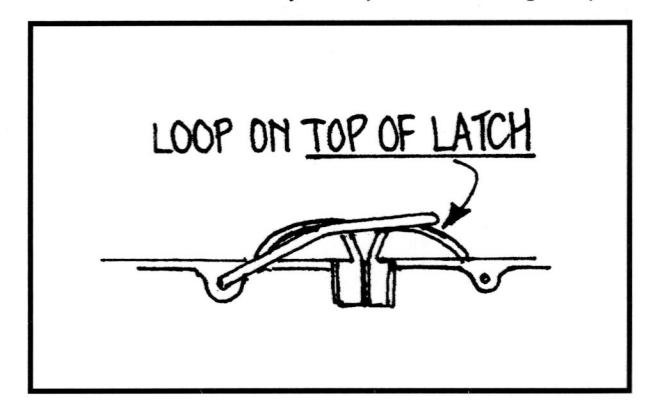

LOOP ON **TOP OF LATCH**

The DGs coupled, with the loop firmly engaged with the hook of the opposing coupling. The actuating dropper can be clearly seen beneath the grain wagon (it could do with adjusting to be a little closer to the track, and the curved delay-latches are also visible, resting on top of the hooks. These couplings don't part unintentionally, even on rough track or tight curves.

tweak it up or down as required) and very forgiving when things are a bit 'out' - very useful at exhibitions. As is so often the case, getting everything working smoothly is a matter of trial and error.

Provided you stick to these two mounting criteria - height and reach - when mounting these couplings, just about any sort of vehicle can be fitted. Locos present no problem, but coaches - especially corridor coaches - can be slightly more complicated. They can, of course, be body-mounted to coaches just as for wagons, but corridor connections will sometimes prevent the loops from rising far enough for the delay feature to work - not usually too much of a problem, as by and large corridor stock doesn't get shunted that much. Curves can also cause problems with these longer vehicles, and it is sometimes necessary to mount the DG's on extension pieces on the bogies rather than on the body.

Operation

Coupling up is simple. The vehicles are brought together and the loops ride up the sloping faces of the hooks, lift the delay latches, and drop behind the hooks. Usually, only one loop drops into place, but this is fine; your vehicles are coupled. Only gentle contact is needed for this to happen. Some modellers - including yours truly - only fit a loop to the coupling at one end of a vehicle. This has no effect whatever on the reliability of operation - in fact, it may improve it slightly - but does have the effect of 'handing' your stock. Turn something round, and you could find yourself loop-less when you come to couple up! If, like me, you have a few vehicles that have had one of those disasters that befall models from time to time - spilt solvent or straying soldering-irons - it's quite hand to have an excuse for always keeping that particular side away from the viewer!

Uncoupling is the same as for other magnetically-operated delayed-action couplers like the Alex Jackson or Sprat and Winkle; the couplings are slackened by setting-back or pushing the stock slowly over the magnet, which is then energised to attract the steel droppers to lift the coupler loops. Continued pushing puts these into the 'delay' position as the droppers pass out of the magnetic field and the loops drop back on top of the delay latches. The vehicles can then be pushed to the final desired location, and when the vehicles are drawn apart the loops travel up the curved backs of the latches to drop clear, leaving the couplings 'primed' for re-coupling when required.

Conclusions

I have found the DG a simple, easy to fit coupling at a very acceptable price. They are extremely reliable in operation, pretty robust, and not too obtrusive. To those who (like me) have struggled with making and setting-up the Jackson, but still hanker after a magnetically-controlled coupling with delay facility, this could be the answer. Don't believe me? Next time you visit an exhibition, look for the layout where the hand in the sky stays in somebody's pocket rather than destroying the illusion; that could well be the layout using DG couplers...

It's not all steam, ma...
If you have a soft spot for Sulzers, a passion for Paxmans or delight in Deltics, then you need the diesel-modeller's vade-mecum:

MODELLING DIESELS
in 4mm Scale
RTR Conversions and Improvements

Tim Shackleton

A RailMODEL Handbook

96 A4 Pages of Shackleton wit and wisdom
190 B & W, 29 colour illustrations
£9.95 post free
RailMODEL PO Box 2 Chagford Devon TQ13 8TZ
Tel/Fax: 01647 433611 - Visa/Mastercard

FOX-TALBOT TALES
The Light Fantastic...

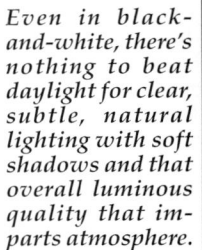

Even in black-and-white, there's nothing to beat daylight for clear, subtle, natural lighting with soft shadows and that overall luminous quality that imparts atmosphere.

In our first look at model railway photography in **Digest 3**, *the subject under scrutiny was the primary basic tool of the job, the camera, with some suggestions as to what might be a good starting point. For this issue, editor* **Iain Rice** *considers the key component in any photograph - the light under which it is taken.*

As any exhibitor at model railway shows is well aware, a considerable number of the visitors to such events evidently wish to record their experience by taking photographs of what they see - an operation all too often involving firing flashguns off right in the operators face! Over the years, I've seen miles of film wasted in this way, for I suspect - nay, I am quite certain - that the very great majority of such photos turn out to be quite useless.

Having, over the years, taken more useless photos than you could shake a stick at, I feel well-qualified to pronounce on the whole business. I too was a 'friend of Kodak' before I finally twigged that, in photography as with all other natural sciences, there are certain fundamentals you simply can't ignore. And enough light of the right sort is most definitely the key fundamental of photography. So - how much light is enough, and what is the right sort?

Adequate Illumination

Capturing a good-quality photographic image depends on the film being used for the job being properly exposed; that is, the total amount of light that reaches it during the taking of the picture must be sufficient to complete the photo-chemical changes needed to create the image, without being enough to overexpose or 'fog' the negative. The amount of light that reaches the film is determined by a number of factors, over most of which we have some control. These include, besides the obvious influence of the brilliance or otherwise of the light source being used to take the photo, the nature of the subject - most particularly, its reflectivity; the 'speed' (light-gathering qualities) of the camera lens and the aperture at which it is being used; the length of the exposure that is practicable in the circumstances; and the sensitivity - or 'speed rating' of the film.

Let's start with the last, as this is very easy to control - you just buy an appropriate type of film. Most film is sold with speed calibrated on the ASA rating; and the higher the ASA number, the more sensitive the film and the less light needed to secure a full exposure. The commonest ratings for modern film stocks - black-and-white, transparency or colour print - are 100, 200 and 400 ASA, (but don't be fooled into thinking these relationships are linear; 400 ASA film isn't four times as fast - in terms of exposure time - as 100ASA). Obviously, where light levels are low, you're more likely to get a result on fast film, but there is a trade-off in terms of the crispness of the image and the amount of enlargement it will take. This is because of the greater particle size of the light-sensitive chemicals in faster films - known as the 'grain' of the film.

For model work, fine-grain (i.e.., slow) film is desirable in the cause of a crisp image that will show fine detail and stand enlargement. This means we need to be able to get more light onto our film to ensure adequate exposure - but that brings problems in other directions. As explained in the camera piece last issue, another prime requirement of effective model photography is a good depth of field, so that as much of the subject as possible is in focus; but good depths of field call for the smallest possible aperture on our lenses - f16 or f22 in most cases - further cutting down the amount of light reaching the film.

All of these factors, taken together, clearly indicate two basic tenets of model photography; you need as much light as you can possibly get to illuminate your subject; and you need the camera firmly supported on a good, solid tripod so that long exposures can be used to sufficiently saturate the film with light to obtain a good image. You're unlikely to get far hand-holding the camera with the lens wide open in a poorly-lit environment.

Why not flash?

I'm not going to claim that you can't take a good model photograph with flash as illumination. My good friend and opposite number in Holland, Len de Vries, gets some stunning results doing just that - but not with some tuppenny-ha'penny little flash-gun built-in to or perched on top of his camera. No, his flash set-up uses multiple sources computer-controlled, with reflectors and fill-flash and composing lights and heaven only knows what else. Quite a different kettle of fish to the sort of equipment most of us have access to.

Flash also has a number of other disadvantages as an approach to photographing models. It is a very intense short-duration light source and unless it is pretty sophisticated (and dedicated or interfaced with the internal metering and electronics system of the camera) it cuts down the number of variables over which you have control - shutter speed and aperture, for a start. Flash lighting also tends to be very harsh, casting strong shadows, while the sort of single-source built-in or perch-on apparatus normal for general use gives a strong mono-directional lighting that is difficult to control and usually far from natural. And, as the flash is instantaneous, there's no way you can view your subject under the same light as that which will illuminate it during the exposure. Add on problems of flare, reflection, synchronisation and the width of the flash-beam (no use having a wide-angle lens on the camera if the flash is a narrow-angle beam!) and flash more or less rules itself out for most purposes.

Suitable Light Sources

Having dispensed with flash, what other suitable light sources are available to us? Basically, two: daylight, and electric floodlighting of various sorts. The latter is a subject in itself, and will be covered in the next edition of the FTT. Which leaves us with daylight - by far the best form of illumination, and totally free to boot!

If at all possible, photographing models outside is always my first preference, as daylight has a quality (and a sufficiency) that no artificial setup will ever match.

Outdoor photography of Arcadia. Here are two sections of the layout perched on the Brent family picnic table, in perilous proximity to the fishpond. (No, I didn't take that one step back too far - but I came perilous close!). Note the improvised set-up to support one leg of the tripod on the seat of the table to allow close-in work.

Daylight, of course, varies with the weather, and can assume many different forms. Given that we're unlikely to be photographing our delicate models outside in the pouring rain, what is the best sort of daylight for our purposes? I've found that the best bet is the range that extends from 'moderate overcast' to 'hazy sunshine'. A dull day is no good for obvious reasons, but brilliant full sun is often equally useless, as you're apt to encounter some of the problems characteristic of flash: strong shadows and glare.

Pitfalls of Photographing under Daylight

There are actually very few drawbacks to photographing models under natural light, apart, possibly, from the difficulty of transporting the model into a suitable location. The photo at the foot of this page shows how Martin Brent and I set 'Arcadia' up in his garden in order to take the pictures gracing this edition of the *Digest*. We photographed the model two-boards-at-a-time, setting them on a picnic table on the terrace. The light was, of course, overhead and coming from the left of the layout and from behind the photographer. One of the main differences between using daylight and artificial light is that you have to move the layout in relation to the light and not vice-versa; so you may need a couple of willing mates to heave the boards about!

The weather on this day was ideal - bright overcast, with occasional bursts of hazy sunshine - and there wasn't too much wind, which is a hazard of outdoor layout photography that shouldn't be underestimated! Fortunately, at this time of year, (September, very pleasant in Herefordshire) the sun was still high enough from mid-morning to mid-afternoon for cast shadows from adjoining buildings, trees and so on not to be a problem; it was also no problem to align the model to make the best of the light, the only hazard being the adjacent fishpond, calling for caution on the part of the photographer stepping back for a 'long view'.

Apart from the tripod and remote release, the only other items needed for outdoor model photography are the tripod and camera remote release, a couple of pieces of white cardboard about 18-ins. square, and a suitable backdrop if the model itself doesn't incorporate one; fortunately, 'Arcadia' does. Mind you, it is sometimes possible to use the 'real' view as a very effective model backdrop - some American photographers, most notably Vic Roseman, have become very adept at blending modelling and reality in this way, to produce some stunningly convincing results.

It seems true that outdoor photos with backgrounds that are, strictly speaking, inappropriate seem generally less distracting than indoor pictures with the same problem - witness Neil Burgess' pictures of his 'Blagdon' layout in this issue. However, these same pictures do give some idea of the problems that stronger sunlight can bring in terms of deep cast shadows, although sometimes, as here, it can produce some dramatic and effective results.

Chris Chapman took this outdoor shot of the tally hut on my 'Tregarrick' layout many years ago; the background is the row of cottages across the street.

Colour Balance
Another advantage of using daylight, especially for colour photography, is that colour-balance problems go away, as all colour film is optimised for this sort of light unless it is of special 'Tungsten' grades. There is no doubt in my mind that colour values in artificially-lit shots rarely if ever look as good as daylight, no matter how clever one is with filters - which may not be surprising, but often seems overlooked.

The differences in black-and-white photography may be more subtle, but they're there none the less; I don't think I could achieve anything like as effective and subtle an illumination as the sun provided for the shot of 'Arcadia' that heads this piece, no matter how long I fussed about with photofloods, reflectors, fill lights and filters. Whereas all I did to get the result shown was to sit the camera (the Contax 139 with 35mm. lens stopped down to f22) actually on the layout, set it to 'auto' and let it get on with it. The same was true of the colour pictures in the Arcadia article.

Fill Lighting Outside
I suppose one of the chief drawbacks one might cite in using available daylight (to give it its arty-title) is that you can't control and direct it like can an electric lamp - and if there's something it won't illuminate, then it's no use expecting the light source to move; although if you hang on a bit it may dim or get brighter of its own volition. However, there are one or two dodges you can employ to get over that common problem, the dark corner.

This can be a bit of a bugbear when photographing locos and stock, where a lot of the detail lies in the underframe - inevitably in deep shadow when the model is lit from overhead, as all outside photos invariably are. There are two simple ways round this: The first is to incline the whole model up on the lit side, effectively lowering the angle of the sun a bit and thus allowing more light to fall into the 'shadow regions'; and the second is to use one of those pieces of white card as a reflector to bounce light back into the shaded areas to help illuminate the detail. It's surprising how effective this is, and how much shadow can be 'killed' with a couple of assistants wielding pieces of card. The card reflector can also be used to light areas like building or loco cab interiors, to reduce the intensity of shadows or to help avoid too-sharp contrasts between lit and unlit surfaces.

Natural Effects
The abundance of light in the great outside often means that you can take photographs quite simply that wouldn't be possible without a great deal of expertise and equipment indoors. Personally, I greatly enjoy playing around with small, simple outdoor photo mock-ups like the one used to produce this month's cover.

The ingredients here, as can be seen from the candid photo below, are very simple: a very crude little diorama consisting of about 15-ins. of track on a low embankment (piece of 12mm. MDFB), with a foreground of hairy carpet felt and other scenic materi-

A simple diorama, a painted backdrop and plenty of nice hazy autumn sunshine were the essential ingredients for this month's cover shot.

RailMODEL Digest

als (construction time oh, the better part of twenty minutes), all set on a card base and a small painted backdrop. Note that the backdrop and the modelled foreground are independent, so that these two components can be adjusted relative to one another. It's a great advantage if you can always keep the backdrop at right-angles to the camera lens - especially if, as here, it has buildings painted on it; get it on the skew, and funny things happen to the perspective!

The final cover shot shows the sort of effects that natural lighting can impart to a model; the sun-dappled boiler, with the shadow of the tree freckling the pannier tank, the clear 'etching' of the silhouette of the dome and chimney against the backdrop (without any awkward sharp shadows, always a problem with photofloods in this situation) and the actual cool autumnal quality of the light. There are some quite strong shadows in this photo, but I preferred this version of the shot to an alternative with light 'bounced' onto the back of the bunker. Apart from revealing a couple of rather wonky lamp-irons, this lessened the 'drama' imparted by the sunlight.

Variety is the Spice...
Proper professional photographers spend a long, long time setting up and checking every aspect of a photo, and click the shutter a couple of times for perfect results. Amateurs like Rice fire off miles of film in the undying hope that somewhere amid the dross will be a few useable shots. This is an approach I would commend to you, as film is cheap and a good variety of alternatives gives you more choice when it comes to selecting the pictures you will finally use.

When making multiple exposures like this, it's important to make sure that you allow for vagaries in the lighting and the reflective nature of the subject by 'bracketing' your shots. This is the practice of exposing one shot at the setting suggested by the camera's metering system, and following this with one a stop over, and possible a further bracket at two stops over. Occasionally, it may also pay to add a further bracket one stop under. (What this means is, for instance, if the camera suggests $1/8$ second as the best exposure, you take one at this setting, one at $1/4$ second and one at $1/2$ second, and possibly one at $1/15$ as well.) This is a good idea because most camera metering systems are weighted to take account of normal conditions, and compensate for a lot of light in the sky, for instance. Hence, a lot of light, reflective colours in a model - the backdrop, perhaps - may 'fool' the meter into thinking there's more light available than is actually reaching a lot of parts of the subject. The more sophisticated - and hence expensive - cameras may have inbuilt systems that can compensate for even the extreme of a dark model against a very light background, but overexposing a stop or two is always a useful insurance policy. Apart from anything else, an overexposed image is almost always useable, as it can be 'taken back' in the printing processes, but an underexposed shot is invariably bin fodder.

The Slide Option
Most of my work is now done on colour slide film, as this is one of the cheapest and most flexible options. I use Jessop's own-brand 100ASA transparency stock, process unpaid, which, at £2.99 a pop for a 36-exposure cassette - even when they're not doing a 'three films for the price of two' offer - is cheap enough. Another £2.99 gets your slides processed unmounted (why pay for someone to mount up slides you may never use? The time to worry about mounts is after you've weeded out the duds!), so £5.98 gets you 36 images, or about 17p a go.

A colour slide can be used to produce a colour print, is the type of original preferred by publishers for colour reproduction, and can easily be scanned as a black-and-white image if needed; a lot of the pictures in the *Digest* start out thus.

Summary
As I hope these notes have elucidated, there's nothing to beat the Good Lord's light when it comes to taking the sort of natural photograph that is the preference of many people. All the best layout photographs I've taken have been taken outside, and no matter whether you're trying to photograph a whole layout or simply a single item of rolling stock, daylight should always be the first choice. At a previous abode, I had an attic with a big north-facing skylight that saved me the cost of buying any photographic lights for several years.

However, it isn't always possible to take layouts outside, or even to move them close to a window, often quite a viable option. In these circumstances, artificial light is needed - and that's a whole different (and regrettably often costly) ball-game; we'll be looking into it in *Digest 5*.

RailMODEL *Digest* Back Numbers

This issue completes our first full volume, so if you've only just discovered us and want to fill in the gaps, then you'll be pleased to know that *Digests* 1, 2 and 3 are still available, as is the special Preview Issue, a few more boxes having surfaced at the printers.
The cost of regular *Digests* is £6.45 an issue; the Preview is £5.95, and we supply all back-numbers - as with all our publications - post free.

We can also now extend our exhibition back-number special offer to mail-order customers for a limited period: Buy issues Preview, 1 and 2, and get £2.50 off Issue 3 a total of just £22.80 for 4 Digests. For those who have the Preview, we are offering Issues 1 and 2 with £1.50 off 3, a total of £17.35. Offer closes 31st. January 1997.

RailMODEL PO Box 2 Chagford Devon TQ13 8TZ Tel/Fax: 01647 433611
Mastercard and Visa mail order available.

After GARLIESTOWN

After where? Well, I suppose that, to the vast majority of Sassenach enthusiasts - and quite a proportion of the Ain Folk as well - Garliestown is not so much insignificant as unheard of. Yet there was a time - a good few years ago now, I'll admit - when Garliestown had a railway that was not just interesting and out of the way - it was probably unique!

For a start, the Garliestown branch of the Wigtowshire Railway was officially a tramway. Or rather, it was supposed to be, as that was what Parliament had authorised - but the Wigtownshire didn't think much of that idea, and felt a proper railway would be a lot more useful. So they carried on and built one, a short (not much over a mile long) but straightforward branch of their slightly wayward 'main line'; it opened on 3rd. April 1876 - it would have been somehow more fitting if they'd kicked of on the first!

Eventually, of course, the Powers that Were caught up with this little irregularity, whereupon the Wigtownshire Railway had to seek powers to abandon a tramway - which it had never built - and to construct the railway which had been there all along. One suspects a certain amount of tongue-in-cheek by the officials involved!

A Caley 'Jumbo' - with original chimney - mainstay of the line in the 1950's. I think this picture was taken at Millisle, but I have no details. A comely engine, and a good subject for a model

Rice Collection/origin unknown..

Curves, Grades, Mills and Steamers

Short it may have been, but the Garliestown branch was not devoid of interest. It managed, in its mile or so, a steep (1 in 58) grade, some flange-grinding reverse curves, a brace of level crossings, a station with an amazingly long curved platform, a big feed mill and a harbour with a passenger service (to the Isle of Man). Indeed, the Garliestown branch boasted its very own brand of boat train, usually an excursion working from a Clyde Valley mill town to the Isle of Man! Some of these boat trains were wild affairs - David L Smith, in his immortal 'Little Railways of South West Scotland' tells of trains so long that they blocked both level crossings, and which took virtually the entire motive power of the railway to get them back over those fierce curves and up that 1 in 58, packed with a steamers-worth of trippers.

80 RailMODEL Digest

The Garliestown branch would be a golden excuse to model one of the most distinctive and characteristic of G & SW locos, the Stirling 0-4-2's of the '221' and succeeding classes. This is 6357 of the '635 class',; half a dozen of her sisters made it into LMS days - just; the last went in 1924.

Rice collection/origin unknown

It's hard, visiting Garliestown these days, to imagine such wild capers, the milling crowds and the overladen steamers racing to get over the harbour bar ahead of the falling tide - it was, apparently, a close-run thing on more than one occasion!. No, the chief relic of the place these days is the big, square, ugly brick-built feed-mill that provided a traffic that kept the line open until 1964. And, though the station site itself may now be given over to the portable suburbia of the caravanisti, it's still easy enough to trace the lines that served the quays and the mill. Shut your eyes and brace yourself against the sea breeze off Wigtown bay, and you can just conjure up a Caley 'Jumbo' with a tin-can lum (Anglice: stovepipe chimney) shuttling grain wagons under the mill loading chutes.

Actually, I've never quite been able to work out why they chose Garliestown, of all places, to stick up a whacking great edifice like this, or quite what its trade was. I suspect that the raw materials of the feed arrived by sea, and I surmise that the end product was for the delectation of the considerable bovine population of this prime dairying area. Nowadays it is (or it was when last I was there, in 1992) a sort of rather Arthur Daley second-hand-and-surplus store - though quite where the customers to keep such a enterprise afloat are to be found in this remote area is quite beyond me!

The Latterday Line

The Garliestown branch lost its passenger service some 46 years ago, in 1950, and - boat trains apart - it never amounted to much when it was running; single-coach trains were the norm. No, it was the goods traffic to the harbour and the mill that kept the line going in its later years, and the tonnages were far from insignificant. There was agricultural traffic and a bit of fish as well as the mill to keep things ticking over, and there was also - I understand - some trade in quarry products through the harbour.

Garliestown itself has little to distinguish it (apart from the only holiday cottage the Rice family have ever given up on!) but it is well situated on the beautiful western shore of Wigtown Bay, with some fine scenery and not a few good sandy beaches close at hand. The climate of this part of Galloway is mild and, in good weather, it's a fine place, especially if you're partial to a spot of ornithology. It is, therefore, not inconceivable that Garliestown *might* have developed a bit as a holiday resort of the minor kind, given a good hotel and a spot of genteel promotion (Garliestown - Jewel of the Machars - you know the sort of thing...)

For the purposes of my little layout scheme here, I have supposed that this was, indeed, the case, and that a fine resort hotel was built on spacious Edwardian lines by the Joint Committee (did I mention that the Wigtownshire Railway came to be managed by a joint committee, the (unlikely) partners in which were the G & SWR, its arch-rival the Caley, and their respective English allies, the Midland and the LNWR? I didn't? Well, it was, until they all got swallowed up by the LMS). To serve this investment, I have also surmised a spot of upgrading of the station and the provision of a few additional facilities for travellers both to the resort and in connection with the Manx steamers.

Garliestown as a model

I have, of course, taken the usual gamut of liberties with this prototype when turning it into a model railway layout design. For a start, in reality the feed mill and harbour are the thick end of a quarter-mile beyond the end of the station platforms, the line running along the side of the road linking town and harbour. So there's quite a lot of compression here, as well as in the size of the feed mill and the generally sprawling nature of the harbour layout. In common with a lot of minor and rural stations, Garliestown sprawls a lot - I reckon it's the better part of a mile from the entry point

The Garliestown goods - might well have provided a refuge (brief, for the LMS were especially ruthless with all ex-G & SW engines) for an old Sou' West 0-6-0, like this handsome Manson seen in its prime.

Rice collection, origin unknown

Garliestown harbour - the feed mill, which was rail-served. Mill and harbour offices in building on left.

Outside staircase detail on the feed mill. I say it's pure 1950's

The office complex, on the other hand, is far older - and far nicer...

The harbour sidings also extended to this warehouse

of the station loop to the bufferstops at the far end of the harbour. So obviously, in trying to condense this straggle onto an L-shaped site manageable within the confines of my shed (internal dimensions 11ft. 6ins. by 7ft. 6ins.) a lot of compromise has been called for.

But the essential geographical relationships between the main elements of Garliestown *have* been preserved - including the kickback lead to the feed mill loading roads - though I've had to cut these down to a single line; there were two on the prototype. This, coupled with the short headshunt to the feed mill (loco plus two grain wagons), makes for interesting shunting moves on the waterfront, with quite a bit of to- and fro-ing to the station runround loop to keep the loco at the right end of the wagons. And no, I don't know why the mill was laid out like this when logic would seem to dictate it should be a mirror-image, with the access to the bins from the station end.

By putting in the extra crossover shown as a broken line on the plan and doing away with the wagon turntables and their connecting siding, it would be possible to make the on-quay workings a good deal more flexible - albeit at some cost in siding spaces and, dare I say it, operating fun. And, of course, if there were a couple of feet more length available, then things could be 'relaxed' a bit, and the second feed mill loading siding reinstated.

Scenery and Structures
There's not a lot to the scenery on this layout - Garliestown is on a more-or-less level site, so a few gentle hillocks, some scrub and a few small trees are about all that is entailed in terms of modelled scenery. What would be needed, though, is a good backdrop; the view across Wigtown Bay towards the Stewartry Shore (Kircudbright) is quite something, and would run down the whole of the long wall where I've marked 'Harbour'.

This may sound daunting, but I think it would actually be fairly straightforward - so long as it was kept subtle. The bay as I remember it usually had a cool blue-silver sheen under a tall sky full of big white clouds, (That was when it wasn't full of big black clouds, accompanied by horizontal rain and a wind that always went the short way - straight through you!), with the hills above Kircudbright blue and dun in the misty distance. All this could be represented very simply, as no detail or perspective would be needed.

The structures are a bit more of a tall order - literally, in the case of the feed mill, which has four and five floors. I've turned the range of buildings that face it - the third picture down on this page - round, so that the long rich frontage will be seen on the model. All these structures could be ideally modelled in card, being just the sort of subjects we had in mind when coming up with the brief for our card modelling series.

Garliestown Trains

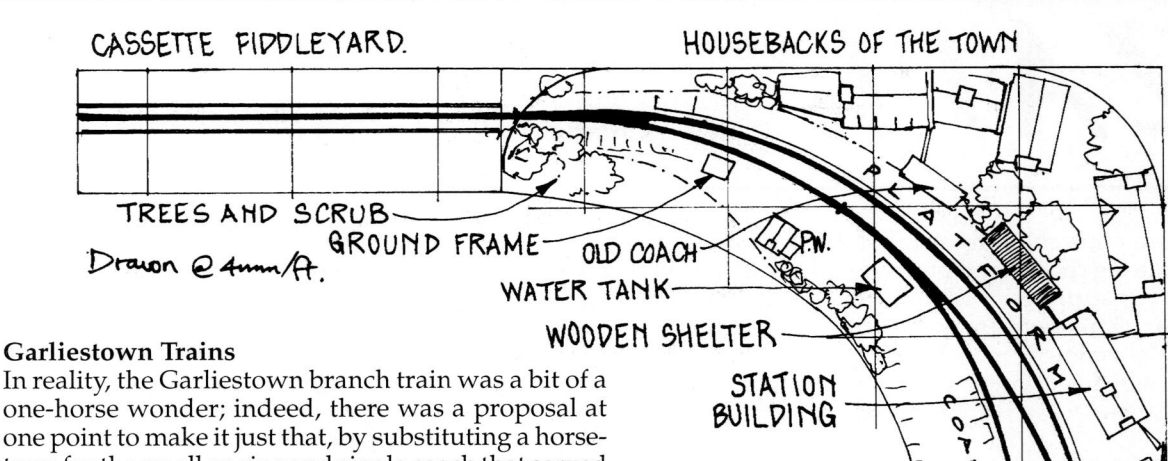

In reality, the Garliestown branch train was a bit of a one-horse wonder; indeed, there was a proposal at one point to make it just that, by substituting a horse-tram for the small engine and single coach that served the line in its earlier days. Due its dubious legality, the line didn't at first have a direct connection to the Wigtownshire, access being by a sort of double-shuffle off a reversing siding, and it wasn't until well after the situation was regularised that a direct connection was made - and even then, this missed the actual station at Millisle, so the Garliestown trains, by and by, went through to the junction with the Carlisle - Stranraer line at Newton Stewart, rather than simply shuttling the mile up to Millisle.

The original trains were, like all Wigtownshire trains, original. The line was run by Thomas Wheatley and his son as a family business, and they were at Colonel Stephens' game on the second-hand market long before he was, in the early 1870's. Latterly, though, things became a little more regular, and in the early-1920's period which I have envisaged as one possible time-setting for the layout, the branch train would have had a distinctly Scottish pre-group look, with ex-Caley and G & SWR stock, hauled by the likes of a Sou' West 221 class 0-4-2 or perhaps a small Caley bogie tank. Goods trains would have been of like ilk - a refuge for an old G & SWR goods loco and a pregroup goods brake. Lovely!

My alternative time-setting would be the later 1950's, supposing the passenger service to have survived a bit longer than it did on account of the holiday trade; so a branch train of a Caley 'Jumbo' - a Rice favourite and mainstay of the Wigtownshire line at this period - plus a couple of Caley coaches would be the thing, with perhaps a MacIntosh shunting tank or Jinty on the goods workings - or even a 4F. Pensioned-off Caley 4-4-0's wouldn't be impossible, either, while for its last few years, the line was the exclusive province of ex-LMS Ivatt class 2 'Micky Mouse' moguls (but not, I fear, the 2-6-2T's).

Of course, with the passenger service dying a decade before Beeching, travellers in Wigtownshire never knew the unparalleled luxury of a BR Mk. 1 non-corridor coach, but we might venture such hedonism on this 'factional' version; who knows, perhaps the passengers would have lasted as long as the cattle-cake, until the line finally closed in 1964.

Summary...

All of which amounts to a branchline layout on a less obvious theme, with a reasonable traffic potential and (to my way of thinking) some jolly nice engines to run it. Slainthe!

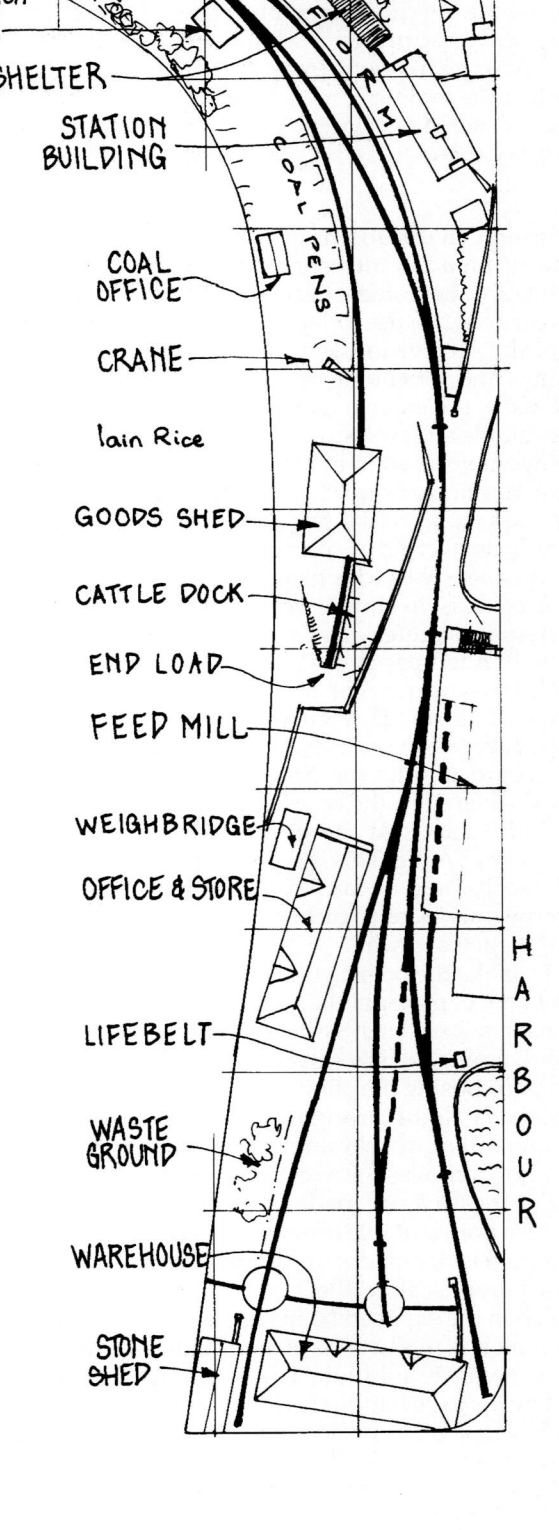

Issue No. 4

ZAP!
Our Regular Electronics Column
Conducted by Dick Ganderton C.Eng. MIEE

In the previous column in this series we looked at light emitting diodes and built a regulator to imrove our power supplies. I though that in this issue I would explain how to make a printed circuit board, look at the magnetic reed switch and say something about where to get circuit ideas and the components to carry them out.

Printed Circuit Boards
The printed circuit board, abbreviated to p.c.b., is probably the commonest form of electronic circuit construction in use today. Just about everything uses a p.c.b. - television sets, computers, washing machines and, of course, model railways. Why? Well, the p.c.b. is easy to make in very large numbers, it is repeatable and well suited for use with the very small components used in modern electronics. It also lends itself to automatic production methods. But, you may well ask, can I make my own boards at home? The simple answer is, of course you can - and you don't need expensive and complex equipment to do so.

A p.c.b. is, in simple terms, an sheet of insulating material - I prefer glass fibre, but s.r.b.p. (synthetic resin bonded paper or Paxolin) is cheaper - covered with thin copper. The track pattern to make up the circuit is formed by chemically etching away the copper between the tracks. Holes are drilled through the board to take the component leads, which are then soldered to the copper 'pads'.

Take a look at the p.c.b. used in the regulator project in ZAP! 3. This is a very simple board to make - in fact it could be made using a sharp scalpel to cut out the track pattern rather than etching it. However, that is the hard way!

To make the board you will need a sheet of single-sided p.c.b. material that has been pre-coated with a photosensitive etch resist. Maplin Electronics sell this in a range of sizes in both s.r.b.p. and glass fibre types. Exposing the photosensitive etch resist to ultra violet (UV) light through a mask changes its composition so that developing it in a sodium hydroxide solution removes the resist where the UV light fell on it. The mask can be made by photocopying the track pattern onto suitable transparent acetate sheet, or by using specially made rub-down 'p.c.b. transfers' onto transparent acetate sheet.

When the exposed board has been developed and washed it is ready to be etched. This is done using a ferric chloride solution. This is extremely corrosive and needs to be handled with care - so beware. A pair of 'Marigolds', to protect your hands, is highly recommended! It is also illegal to wash it down the sink!

I use a plastics ice cream tub as the etching tank. The embryo p.c.b. is immersed in the etching solution and left. To help in getting the board into and out of the etching tub I make a 'sling' out of a length of masking tape across the non-copper side of the board. This allows me to manouever the board in or out of the etching solution and rock it without having to put my hands into the solution, rubber gloves or no rubber gloves!

Etching takes place faster in a warm solution and faster still if the board is rocked or the etchant agitated. Bubbling air upwards over the board by using a fish tank air pump, speeds up things nicely as well. Watch the copper as it is etched away and as soon as all the exposed copper has disappeared, carefully remove the board from the etching tank and put it into another tub of clean water to wash any remaining etchant away it.

To remove the unwanted resist the board is exposed to the UV light without the mask and developed until all the resist is washed off. The board can now be cut to size and 1mm. diameter holes drilled for the component leads.

To finish the board, I scrub it with some industrial Scotchbrite. This ensures that the copper is clean and can be easily soldered to. Professionally made printed circuit boards are usually 'roller tinned', but this is a process beyond the means of most hobbyists.

Veroboard
An alternative to using a p.c.b. is Veroboard. This is an SRBP sheet with parallel strips of copper on one side. A matrix of holes is pre-punched in the board at the standard pitch of 0.010 ins. By cutting the tracks at appropriate places and linking across tracks with tinned copper wire and electronic components, complete circuits can be built in a similar manner to a p.c.b.

This technique was originally developed by Vero Electronics - hence the name - for use in design and development work. It has the advantage of being very easy to use and readily available almost anywhere in the world. As a consequence of this simplicity, it is widely used by the hobbyist and many electronics and radio magazine projects have used the system.

Matrix board is similar to Veroboard, but without the copper strips. The components are inserted in the holes and their leads joined together by soldering tinned copper wire to form the circuit. Obviously, the tinned copper wire needs to be carefully routed to avoid any unwanted short circuits. It is possible to simulate a conventional printed circuit board with reasonable accuracy to check it's viability before progressing to a final conventional p.c.b.

Magnetic Reed Switches
There is one type of switch that is different in operation and construction to all other switches, and it is very useful to railway modellers. The reed switch consists of a pair of special contacts made from a magnetic material sealed in a glass envelope (**Fig. 1**). The contacts are normally held open by the springi-

ness of the contact blades, but when a magnetic field is present one of the contacts becomes a 'north' pole, the other a 'south' pole. Like magnetic poles attract each other, so the contacts snap close. The magnetic field can be generated by a permanent bar magnet or by a suitable coil wound around the reed switch body.

Reed switches are available in different sizes, the smallest - a mere 3.2mm. diameter x 20.3mm. long (although the leads make the total length 57mm.) being probably the most useful to us. They are also available with the coil in a small plastics package, rather like a standard dual in-line (d.i.l.) integrated circuit, as a 'reed relay'.

Like all switches, reed switches have maximum voltage and current ratings. For the smallest (Maplin Part No. FX70M) the maximum current the contacts are capable of carrying is 500mA (0.5A) a.c. or d.c., while the maximum switching voltage is 200V d.c. They have a 'life' of 100 million operations and operate in 1ms and open in 200μs - that's fast for a switch!

If you keep your eyes open reed switches can be picked up very cheaply indeed. I have just had a look around the various stands at an amateur radio rally near Chester-le-Street and found a couple of stands selling them at 25p. each. I thought that was very reasonable - Maplin's price is 70p. each for their Part No. FX70M - until I came across another stand with a large box of the things at the staggering price of 6p. each! If you cannot get along to an amateur radio rally try looking at the various advertisements for companies specialising in components, either new or surplus, in the electronics magazines.

Uses

What can we use a reed switch for? I can think of several ideas - from operating a whistle or diesel horn to signalling and level crossing gates. With a miniature reed switch buried in the ballast between the rails, any vehicle fitted with a small bar magnet under the floor, with its axis aligned along the length of the vehicle, will close the contacts as the vehicle passes over the switch and so activate whatever is connected to the contacts.

As well as operating the reed switch by bringing a magnet close to it, or energising a coil wound round the switch body, it is also possible to leave a magnet permanently alongside the switch - so that the contacts are closed - and arrange to open them by passing a sheet of steel between the magnet and the reed switch. This opens up several other possibilities.

By attaching a thin sheet of steel - or other magnetic material, such as suitably shaped pieces cut from old transformer laminations - to a suitable moving part of a mechanism and cutting an aperture in the sheet so that when the mechanism is at the point in its travel where the reed switch is required to close, the aperture uncovers the magnet, the reed switch closes and operates an auxiliary circuit. This auxiliary circuit could be an indicator lamp, a signal, or another motor driven mechanism.

Of course, the steel sheet could be arranged so that as the mechanism reaches the end of its travel the sheet shields the magnet from the switch and turns the mechanism off. In fact, the uses of reed switches on a layout is only limited by your imagination - how about using such a system to stop a turntable at the appropriate positions? I think we'll take a look at the different ways of turntable positioning in a future ZAP!

Practical Considerations

Reed switches are available in the normally open form. Although it is possible to obtain a changeover version, these are not cheap (Maplin prices about £2 each) and are not readily obtained on the 'surplus' market. Do not be tempted to crop the ends of the switch leads to save space. The leads form a vital part of the magnetic circuit and you could seriously alter the characteristics of the switch. You could also break the metal to glass seals and so destroy the switch.

Overload Protection

One idea that occurred to me some time ago was to use the miniature reed switch as a coreless motor protection device - instead of a fuse. By winding a suitable number of turns of relatively thick enamelled copper wire around the reed switch body the contacts will close when the current through the coil reaches a certain level.

It is possible to calculate the number of turns needed, but I did it by trial and error - or as the posher lot say, empirically! Wind 100 turns of 1.25mm diameter (18 s.w.g.) enamelled copper wire round the glass body of the switch. Use your multi-meter, set to the 10V d.c. range, with the simple test circuit shown in **Fig. 2** to measure the voltage across the 18Ω limiting resistor, and so by using Ohm's Law, the current flowing in the coil at the instant the contacts close ($I = V/R$). This will give you the basis for deciding how many turns to wind on the coil for the maximum current you want your motor to draw before the protection comes into operation.

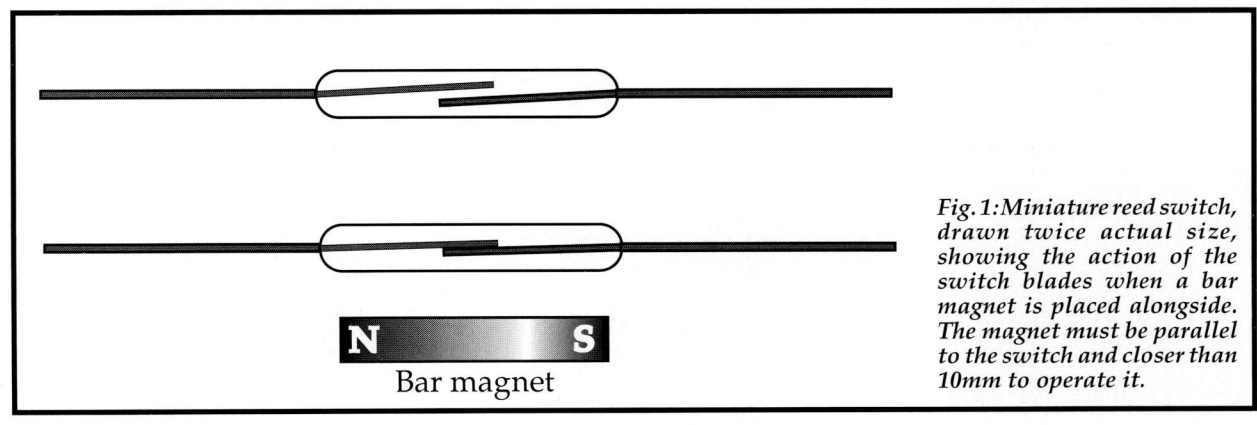

Fig. 1: Miniature reed switch, drawn twice actual size, showing the action of the switch blades when a bar magnet is placed alongside. The magnet must be parallel to the switch and closer than 10mm to operate it.

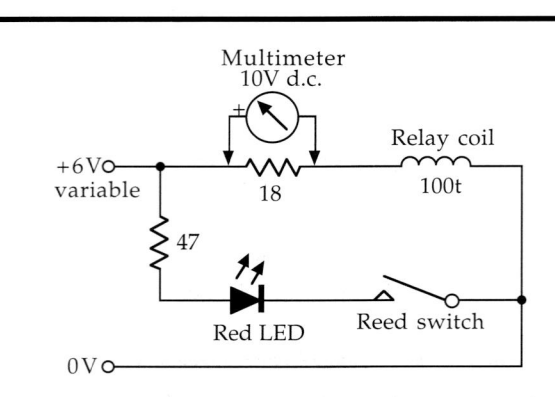

Fig. 2: Simple test circuit to determine the sensitivity of a reed switch.

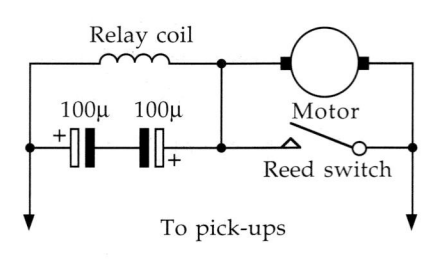

Fig. 3: Circuit diagram of the reed switch overcurrent protection unit.

The red l.e.d. indicates when the switch is closed. Note the voltage across the resistor when the l.e.d. comes on, then reduce the voltage until the l.e.d. goes out. Write the two values of current - the 'pull-in' and 'drop-out' values - on a length of masking tape wound over the coil to protect it from damage.

Let's suppose that 350mA is the maximum current you want to allow through the motor, and that with 100 turns wound on the reed switch body the switch operates at 300mA (5.4V on the multimeter). This gives us a basic sensitivity of 300 x 1000 = 30000mA turns (30A turns). So, for the switch to operate at a coil current of 350mA, the coil will need to have 30000/350 = 86 turns wound on it.

Ideally the test circuit requires a variable voltage power supply to enable the operating point of the switch to be determined - that's yet another constructional project for a future ZAP! session.

The wire used for the coil needs to be as thick as possible because the coil is wired in series with the motor and the coil resistance needs to be a low as possible so as to minimise any adverse effects on the control of the motor.

The switch contacts are wired in parallel with (across) the motor so that when the switch is operated by the motor current rising above the maximum level determined by the coil, the motor is shorted out by the switch and the controller overload protection brought into operation. This also tends to stop the motor dead as the switch shorts the motor, but at least the motor is protected against excess current. The unit is small enough to fit inside a 7mm. scale loco and replace the fuse suggested to protect motors such as the Portescap RG7.

If you find that the switch operates too quickly and trips on short duration current surges, such as occur when starting with a load, try 'slugging' the coil with a pair of miniature 100μF 16V electrolytic capacitors connected in series opposing - positive to positive, or negative to negative, it doesn't matter which - across the coil.

Detectors

I was going to look into some of the different ways of detecting the presence of a train on the track, but I haven't left myself enough room - so it will have to wait until the next ZAP!

Books

With modelling railways you can never have enough prototype information and electronics is very much the same - you can never have enough information. There are many books available on all the different aspects of the subject, but I will point you in the direction of some that I can recommend.

Component Catalogues: You have to pay for these, at least for those worth having! You need at least the *Maplin Catalogue* - now called MPS - and available from W H Smith at £3.45. This 1176-page tome contains a lot of useful technical information and it also has £50 worth of vouchers for use with your orders. Also very useful is the *Electromail Catalogue*. (Electromail, PO Box 33 Corby, Northants NN17 9EL Tel: (01536)204555). This is the same as the RS Components Catalogue and a very useful reference work.

For Beginners: Bernard Babani produce a range of inexpensive paper-backs dealing with just about every aspect of electronics. These tend to have alluring titles such as 50 *Simple L.E.D. Circuits* or *Electronic Timer Projects*.

The Complete Book of Model Railway Electronics by Roger Amos is a book of circuits, ideas and information, all very relevent to model railways.

34 New Electronic Projects for Model Railroaders by Dr Peter J. Thorne is an American book with a very practical approach to the subject. The projects range from very simple to relatively complex.

Send me an E-mail or letter, with a stamped addressed envelope, via the Chagford Potting Shed, and I will send you a list of useful titles.

As far as I'm aware, the electrons have still not reached the outpost known as the Chagford Potting-Shed, so we cannot, as yet, send E-mail to Editor Rice.*(They hadn't by copy-deadline time, but Telecom and the Gods willing, will be in residence in time for - well, Christmas, we suppose - Eds.)*

However, *I* can be contacted via this marvel of modern technology, and I really would welcome your E-mails, or letters, with suggestions for topics to be covered in ZAP!, circuits or problems.

My own personal E-mail address is: **dick-graskop@bournemouth-net.co.uk** - I've had a few communications now, but please send me some messages so that I can get some idea of how many of you are connected.

S & T

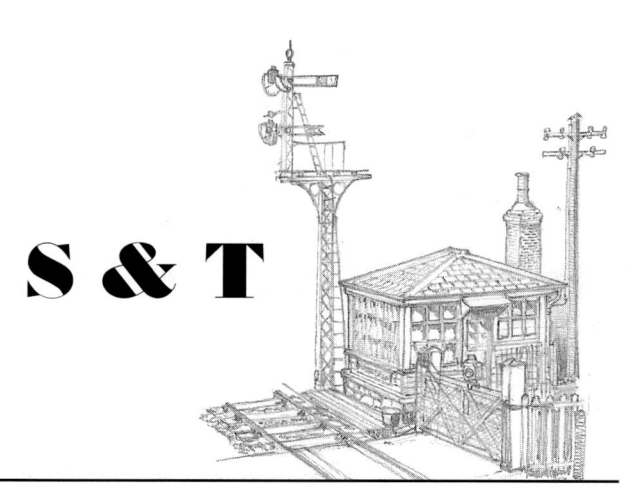

A brief look in for the S & T department this issue, with a couple of unusual signals from widely different parts of the British Isles; is there no end to the fascination and variety of British railway signalling?

Right: The period immediately before and after the first world war saw a number of railways experimenting with new materials for various items of equipment. Prominent amongst these was the search for an alternative to wood for things like signal posts - but without the cost and complexity of lattice. The concrete-casting experiments carried out by William Marriot on the M & GN Jt. are well-known, but what is less often appreciated is that he wasn't alone; the Furness Railway, too, tried concrete signal posts, like this example (with Saxby & Farmer fittings) at Wraysholme Crossing.

RailMODEL Collection, origin unknown.

This is a much better-known - nay, a famous signal, the down home at Midford, S & D Jt. Rly. The first question most people ask on seeing this picture is - where's the railway? The answer is - under the photographer's feet, for this unusual signal is sited on top of 'Bridge No. 17' of the S & D - which took the form of a short tunnel some 25 yards long where a minor road crosses the line at a very acute angle. The lower arm is a backing arm, one of two in the 16-lever Midford frame; they were used to enable trains that had stalled on the bank between Midford and Coombe Down tunnel to set back onto the Up line on Midford Viaduct. There was a special telephone at the tunnel entrance to enable this procedure to be initiated by the train crew of any hapless train that failed to make the grade.

The signal itself is an interesting mix, with an LSWR-style lattice post and original LQ arms, modernised by the installation of power actuation. Well, given its location perched atop a mini-tunnel some twenty feet above the PW, it could scarcely have been easy to arrange conventional pull-wires to the signal cabin, although in earlier times they must have done so. There is an associated relay cabinet at the foot of the signal, which, odder and odder, is sited on the verge of the road furtherest away from oncoming trains - although this may well be a relic of earlier cable operation; the signal-wires could scarcely have been carried across the highway!

Model signalling - where there is any - often tends to the mundane; it would be nice to see some of the less-usual installations of the type featured here being reproduced from time to time.

Rice Collection/unknown

T.P.O.

SR 20-Ton Mineral Wagons, from Mr. Alex Hodson

Sirs,
Reference the article in 'C & W Dept' in *Digest* 2 on the SR 20T mineral wagons - in the introductory note you state that: 'SR freight stock (*goods* stock, surely, at that date?) was generally on a par with, or even in advance of, that of the other companies.'

I feel that this statement needs putting into context. The L & YR - and probably other companies - were building wagons of equal or greater capacity some thirty years before the SR 20-tonners appeared. Consider, for example, the LYR Dia. 56 20-ton 5-plank fruit open - long wheelbase, vacuum fitted, although admittedly still with grease axleboxes - delivered in 1903. A year earlier the L & Y built a 45-ton bogie goods wagon, vacuum fitted and with new-pattern oil boxes, as well as a 30-ton covered van to the same general specification.

When the LYR amalgamated with (or took over, as we tend to say at Horwich) the LNWR, 20% of its entire freight stock was vacuum fitted, more than any of the other pre-group companies and, I suspect, a far higher proportion than on the SR in 1933. A bit *below* par, perhaps?

Alex Hodson, Launceston, Cornwall

M & GN Matters, from Mr. Nigel Digby

Sirs,
Thank you for your splendid promotional job on the 'joint' in *Digest 1*; there were a few points I thought needed clarification.

On closure dates, the Joint was almost entirely closed to passengers in February 1959, with only the Melton - Cromer section open until 1964; other isolated parts remained for goods traffic until 1965 - 8. The last train ran on the Lenwade spur in 1983, while the Cromer - Sheringham section remains BR property. Thus, most of the Joint was closed *before* Beeching.

Some minor points from the text- you also used the old spelling 'Sherringham' - not used after (I think) 1896. *(I was using a rather old map of my grandfathers', whence Sherringham had both its 'r's - although I don't think it was pre-1896. I.A.R.)* There were only two swing bridges on the M & GN, not the several you imply. Also, I don't consider the Norfolk landscape flat at all except in fenland and the Broads. Try a cycling holiday over it *(I have - it isn't. But I didn't think I said it was. I. A. R.)* I'd also question your assertion that diamond fencing is necessarily a Midland design - it was quite widespread.

There are a few points on the photo-captions as well. Firstly, I don't think that the picture of E & M 4-4-0 No. 34 is as late as 1896 - c. 1890 is more like it. The chocolate livery was not as dark as is often suggested.

The 'cottage' station is Lenwade, not Drayton, and that is M & GN trellis fencing, not MR-pattern. The photo is probably by E. Tuddenham, now forming part of the M & GN circle collection.

The date of the South Lynn photo is at least 1930, and the goods in the picture beneath is running *into* Melton, not leaving. The loco is in M & GN dark brown, with painted numerals replacing the brass originals. The M & GN had several ballast brakes, and two of these, Nos. 13 and 101, are shown.

The train behind the Beyer-Peacock 4-4-0 on P.47 is composed of a mixture of ex-E & M 'small' stock and ex-MR coaches. In fact, judging by the headcode disc, this may still be E & M period. I'd also query your assertion about a 'good number' of loco lettering styles - there were two main ones, plus a couple of short-lived early variants and the arched lettering unique to the 4-4-2T's.

In the lower picture of 4-4-0 No. 22, I'd say this is still the ochre livery, which always came out dark on photo plates - the small lettering is the giveaway. The 'deep green' mentioned is a red herring and completely unsubstantiated. *(But mentioned in some published books on the M & GN, nevertheless - I.A.R.)*

In the Sutton Bridge piece, that's a MR-pattern signal being passed by 4-4-0 No. 28, as is the signal behind reboilered 0-6-0 84; the LNER put a little boiler back on her again.

In Pre-Group Glory, the colour plate of No. 18 shows her with extended smokebox c1910, not in original condition as stated. The express is actually a Cromer - Kings Cross working.

Lastly, I was (naturally) a bit disappointed to find that my book 'A Guide to the M & GN' was missing from the bibliography - although this doesn't surprise me, as some shops find it hard to get hold of. It isn't really!

Nigel Digby, Cromer, Norfolk.

Locomotive Studies, from Mr. Eddie Castellan

Sirs,
By way of fine-tuning the excellent 'Prototype Study' articles, it would be invaluable to have a cab interior photo/drawing wherever possible - you can't build a serious fine scale engine without this kind of detail these days! Pictures from odd angles - tender tops, backs, close-ups of fittings and other details are also invaluable.

I make these requests in the full knowledge that it can be very awkward to find this type of picture, especially to publish. But if such pictures do exist - in other publications, perhaps - how about references to their whereabouts?

Also, if you're preparing drawings from a library works drawing or similar source, why not state the source and reference number? That way, we masochists can get a copy if we need one!

Mike Sharman's piece on locomotive drawings raised some interesting points. He didn't mention one dodge that can work well if you've got a good broadside view taken square on - such as a works photo - you can photocopy it to the correct size for your model, and use parts of the photocopy as templates for cutting out components.

I first did this with a reach-rod of a very awkward shape; I knew the distance between the two pin-joint holes, so I copied the photo to size, cut the reach-rod out, glued it to a piece of metal and piercing-sawed round it. Crude? Well, yes - but I ended up with a rod exactly the same shape as the one in the photo! *(A good dodge, this - but do beware parallax error, prevalent on these old pictures. This tends to stretch things a bit towards the edge of the image - so while dimensions in the centre of the photo may be true, the extremities are likely to be distorted; I know all about this - it's how I came to make my built-from-photos GER T7 0-4-2T a scale 18-ins. too long - I. A. R.)*

I'm looking forward to more photographic information - especially as to how I can take a three-quarter front view of one of my 7mm. locos and have enough depth of field to get the whole thing in focus. A 50mm. lens stopped down to F16 won't do it!

Lastly, while I applaud your decision not to publish 'fan mail', I'd urge you to be careful that you don't only end up with the whingeing letters about missing rivets; they don't make very lively reading!

Eddie Castellan, Stafford.

Modelling Wartime Railways, from Mr. Tony Cane.

Sirs,
Congratulations on producing a bumper issue of the *Digest* - 112 pages! No, I haven't miscounted, I got the first colour section and a few other pages twice! *(OK, own up - who's got the one with the first colour section missing? And Tony - we make that another 49.6p. you owe us. Ta. - Eds.)*

My *real* reason for writing is to add some follow-up information to the Pillbox article. I've included Xerox copies of a couple of prints from the Imperial War Museum collection *(Sorry we can't reproduce these here - we'll try and get prints and permission for a future issue - Eds.)* which show railway-related disguised pillboxes, one being painted to look like a wagon (lettered 'LNE' - was any *real* wagon so lettered?) and the other being disguised as stopblocks at the end of a pair of sidings.

The 'wagon' pillbox may not look very convincing in the photo, but put it 100 yards away in the company of other wagons and it is very effective camouflage, as another picture in the I.W.M. collection shows. It's interesting to note that they have gone to the trouble of painting a representation of the wheels and brake gear on the lower part.

There has been published what must be the definitive book on British pillboxes. It is entitled: Pillboxes; A Study of UK Defences, 1940', by Henry Wills, published by Leo Cooper in 1985, ISBN 0 436571 60 1. This includes scale plans for 14 types of pillbox and dozens of photos - including six of boxes on railway property - plus details of the camouflage schemes. The information was collected through local newspapers, and O.S. grid references of all reported surviving pillboxes are given.

I picked my copy up remaindered in a cheap book shop five years ago, so I presume it is now long out of print. But prior to this, my local library was able to supply a copy.

The caption in the article referring to a brick-built pillbox is only partially accurate. The large numbers of boxes being built in the early part of the war meant that timber shuttering was in such short supply that bricks had to be used for this purpose. There being no need to remove the bricks after construction was complete, the illusion of a brick structure was preserved.

On my WW2 period 4mm. scale layout I have, of course, a pillbox, plus a set of tank traps, also covered by the Wills book. Research into, and modelling of, Britain's railways during WW2 is a much-neglected subject, but this is something that I and fellow-members of the World War Two Railway Study Group are trying to put right - our prospectus is enclosed. There are currently about 70 of us, with a good mix of modellers and historians. My own responsibility is for the database of photos and drawings; the photo list is now over 6000 items, so there is plenty of data if one knows where to look. The membership secretary is : Mike Christiansen, 25 Woodcote Road, Leamington Spa, Warwickshire, CV23 6PZ. Tel: 01926 429378.

Tony Cane, Stanwell, Middlesex.

Editors note: We are already working on a couple of WW2 related projects, including a prototype/modelling feature on the US Army Transportation Corps wagons used both here and in Europe. More on this soon!

Photographic Notes, from Mr. Alan Williamson.

Sirs,
Your series on photography, while generally sound in basics, contained a couple of errors of fact. Firstly, Leica cameras use their own Leitz-made lenses, although I for one would not argue if you are contending that Zeiss lenses are as good!

I'm also not quite sure why you think the Contax is a better tool for the model-photography job than other top-flight SLR's *(The reasons given by the late A. E. 'Tony' Smith in his advice to me, Barry Norman et al was the combination of the Zeiss lenses and the metering system which allows automatic time exposures greater than 1 second - I. A. R.).*

I would also point out that Canon cameras only have one 'n' - if it's got two 'n's you get into trouble for shooting things with it! Also, Canon have always used a bayonet-fix lens, but I'd agree that old Canon, Pentax, Yashica or Nikon lenses are worth pursuing. Whether they're as cheap as you claim I'm not so sure. Nikon gear is never cheap, and some of this older stuff is now starting to become collectible.

I'd be very interested if, at some future point in this series, you might come on to consider twin-lens reflex cameras like the Rollieflex, which were at one time much-used for model photography. Also, what about the pin-hole attachments sold in the US for many SLR's; just the thing for this job, surely?

Alan Williamson, Leeds.

REVIEWS
Wychbury/Paul Bellon MOD Hunslett 0-6-0ST, HO Scale; : Eileen's Emporium Flux; Backwoods Miniatures Gearbox; Book: British Railway Wagons: Opens and Hoppers.

Anglo-Belgian kit for Dutch engine? - Must be the EC!

Wychbury/Bellon HO Hunslet

Wychbury Models all-etched kit version of the ubiquitous Hunslett MOD 'Austerity' 0-6-0ST is now available in all scales from 10mm./ft. Gauge 1 down to 3.5mm./ft. HO - the first loco kit to have such wide coverage? The 'base' artwork is apparently the 7mm. version which appeared first, closely followed by a 4mm. derivative.

This 3.5mm. scale model is a collaborative venture with Belgian kit-seller and manufacturer Paul Bellon, who has an interesting HO range including etched brass coach sides for Belgian and, I think, French steel-sided stock, a cast whitemetal kit for the SNCB 51 class Belpaire 0-6-2PT and now a further etched kit, also Wychbury-sourced, for the '18' class, one of the 'Scottish' 4-4-0 types ordered by the Belgian State Railways to Caledonian Railway 'Dunalastair' designs early in the present century.

Paul was interested in the MOD 0-6-0ST because these engines saw much service in the low countries during the latter part of the last war, and quite a number were taken over by the Nederlands Spoorwegen in about 1946 as the '8800' series. It is these Dutch engines that the kit is principally aimed at, but parts are included to make the British industrial and BR J94 variants, making the kit a valuable asset to the growing band of British HO modellers.

The Dutch 8800's were basically the stock MOD design unaltered. They did not have the centre steps of some variants, kept the low plain bunker, standard cab arrangements and fittings. The only 'extras' are Dutch lamp-irons for the usual large oil headlamps, and a few different handrails.

The Wychbury Kit

Like most of Wychbury's range, this is a pretty well-thought-out kit, although there are one or two aspects that could be better. The model has an etched fold-up chassis in 18thou. nickel-silver, which makes token provision for compensation (half-etched guides for hornblock cut-outs). Additional L-section frame spacers, in addition to the integral fold-down arrangements, make a very square and rigid chassis possible - a very sound arrangement.

The body is in 12thou. brass, of a softish grade, with the saddle tank shell and boiler bottom preformed to shape. All the fittings are cast in whitemetal. These include cab fittings and injectors as well as the basic boiler mountings, the castings being quite acceptable if not of the highest quality. The only compromise seems to be the buffers, which are 4mm. items adapted - giving an appropriate head size but rather-too-hefty shanks and housings.

The instructions would appear to be similar to those from the 4mm. kit and are perfectly adequate, with clear exploded diagrams to aid assembly. I suspect that the advice that a Mashima 1220 motor and 38:1 Ultrascale gears, used with the (rather poor) mount supplied, 'can be fitted in the firebox' also comes verbatim from the 4mm. version; no way will a 1220 Mashima fit in the firebox of this HO model, and even the 9-16 open frame motor I used called for quite a bit of hacking to obtain clearance, although this may have been made slightly worse by my decision to use larger-diameter 50:1 gears with a home-made mount rather than the 38:1 suggested.

Comparison of the Wychbury/Bellon HO model with the popular 4mm. scale version by Dapol is telling. HO is a lot smaller. Character of prototype well caught by kit. Marker lamp brackets on rear bunker corners are Dutch fitting as, of course, is outsize oil headlamp. Kit includes alternative hopper bunker/cab for UK J94 and industrial versions.

These caveats aside, the Wychbury kit is a good example of mid-range British practice, and was neither difficult to construct nor costly to complete. I opted for the 'compensated' variant, using London Road Models cast-brass hornblocks, while for wheels for my P87 version I obtained some 15mm. diamer 13-spoke PB - just right - wheels in P4 profile from the ever-useful Sharman 'millimetre' range.

Chassis

The chassis went together very well, and I followed absolutely standard Flexi-chas principles in setting up the hornblocks using the coupling rods and a set of LRM jig-axles. The instructions suggest top-mounted wiper pick-ups, but I found there was enough room for less-obtrusive arrangements on the bottom of the frames. The Mashim 9 -16 open frame motor was mounted to a simple plate soldered into the chassis, and the Ultrascale 50:1 gearwheel was fixed centrally on the fixed rear driven axle with Loctite 901.

Superstructure

As with most model locos, this is based around the footplate. On these Hunslet engines, this is a very simple assembly, with shallow footplate angles and deep, hefty bufferbeams. The footplate angles are simply thin etched strips which are soldered into half-etched grooves beneath the footplate; not ideal, but fairly standard practice.

However, as is so often the case with etched kits, thoughtlessly-placed attachment nibs on these thin strip etches were very hard to remove, yet it is essential at least one edge be clean and true if the angles are to sit into the groove properly without distorting the footplate. Along with nibs on the inside of convex curves, tabs on the sides of thin components are a particular bete noir chez Rice, the more so as they are very rarely necessary!

The rest of the superstructure went together without significant problems, even the arrangement for joining the two sections of the saddle tank - about which I will confess myself a bit dubious when I first set eyes on it! As already noted, it was necessary to create extra clearance for the motor, in the shape of a rectangular cut-out in the boiler bottom forward of the firebox. This was accomplished with an abrasive slitting disc in a mini-drill after the tank/boiler/firebox assembly had been solidly soldered up.

The pre-formed saddle tank was a pretty good fit on the tank ends, but the boiler needed rolling to a tighter radius, and also trimming slightly for length. To complete the saddle tank sub-assembly, there are quite a few handrails, etc., to fit, not all of which could be soldered from the inside of the tank due to the limited opening. A resistance soldering system was useful in attaching these components. The cast fittings were glued in place with Araldite Rapid. I added about 80 grams of lead ballast inside the saddle tank for adhesion.

The cab/bunker is the only other assembly, and here it proved tricky to form the curved eaves that start about 0.75mm. above the cab doorway without the curve 'migrating' to the top of the door opening. Maybe a one-piece cabsides/roof etching would be easier to form? The cab interior is completely clear of mechanism, so the cast backhead and reverser supplied could be fitted without poroblem.

Finishing

As I was finishing my model as a Dutch NS 8800 class, I fitted Dutch lamp-irons and (scratchbuilt) oil headlamps. I also replaced some of the cast pipework on the injector with wire (there's also a 'flat' etched injector on the fret, not *really* a good idea!) and added a few twiddles in the way of lubricator pipes and small grabs. Paintwork for this version will be plain dark green - Panzer Corps Green is about right, they tell me...

Conclusions

Building a prototype with which one is thoroughly familiar in an unfamiliar scale proved an interesting exercise. There is really nothing to complain at in this kit, which goes together well, looks like its prototype and is reasonably priced. Indeed, in comparison with similar HO etched-brass kits of European origin, it's in the 'free with cornflakes' category.

HO is a very useful sace-saver over 4mm., and this loco opens up intriguing possibilities for minimum-space industrial layouts and the like, especially now that wagon kits and wagon parts are becoming available from the British 1:87 Society. Wychbury have anounced other HO releases, including that Caley 4-4-0 in British format and the Fowler LMS 0-6-0 dock tank, another very useful minimum-space prototype.

Iain Rice

Eileen's Emporium - Frys Powerflow Neutral Paste Flux.

Flux to most people is, well - just flux. But not so, as those who may have read my writings on the whole business of flux and solders over the past few years may be aware. Actually, it's pretty vital stuff, as it not only determines the facility with which joints can be made, but also their ultimate strength and life. Life? Well, yes - as those of us engaged in restoring old model locomotives for the Model Railway Heritage Trust are finding, soldered joints can deteriorate quite dramatically over time, due to chemical changes - effectively corrosion - taking place in the joint long after the soldering has been completed.

Traditionally, we attempt to remove or at least neutralise the remains of any flux, paste or liquid, that we have used in the construction of our models. To be effective, all such traditional materials have, by their very nature, been corrosive to some degree or other, some of them very considerably so, making this an important aspect of model finishing. In addition, many of the resin pastes - such as 'Fluxite' and the core flux in Multicore solders - tend to leave burnt residues that look unsightly and impede painting.

It has also been true in the past that the more effective a flux, the more it seemed to produce these unwanted and potentially damaging side effects. So a flux which is effectively chemically neutral once heated, and which leaves no appreciable residue, has got to be good news. And that is what this new Fry's product, 'Powerflow Flux', being marketed for model railway purposes by Eileen's Emporium jointly with JPL Models, claims to achieve. It is a paste, but far removed from the traditional resins, resembling nothing more than Mrs. Rice's foundation cream. However, given that the pot carries warnings to avoid skin and eye contact, I doubt it would do much for the complexion.

What is certain is that it does a lot for the swift, sure making of soldered joints. Eileen's Jim was adamant that once we'd tried it we would be pitching all other fluxes in the bin. Well, I'm not sure about that, but it certainly is very effective, especially for track-building - it works a dream on PCB sleepering and steel rail - and would also seem a pretty good general purpose flux for most modelling and wiring jobs - although there are some things for which I still prefer a drop of the old Phosphoric.

The stuff is formulated for use with lead-free solders, is 'water friendly' in plumbing terms, and calls for no more of a post-joining clean-up than a swift wipe with a damp tissue. It seems to cut through a lot of crud and general detrius, works well with most types of modelling solder and doesn't foul up the bit of your soldering iron. In short, it's an excellent all-round general-purpose material, and at £2.25 for a 50g. bright yellow unbreakable, hard to tip over screw-top plastic pot - about 5 years supply for the average modeller - it won't break the bank, either. Good stuff.

Iain Rice

Frys Powerflow Superflux:
Eileen's Emporium, 55 Reedsdale Gardens, Gildersome, Leeds, West Yorkshire, LS 27 7JD Tel: 01532 537347
JPL, Unit 12, Tyldesley House' Elliott Street, Tyldesley, Manchester M29 8DS
Tel: 01942 896138

High Level/Backwoods Miniatures 4mm. Scale Reduction Gearboxes

Small locos, in 4mm at any rate, usually mean small motors and small motors - certainly when used with conventional gears of the order of 40:1 or so - mean low revs and consequently poor low-speed performance. The problem is exacerbated by extremes of wheel diameter - say, 3ft 6ins or below or 6ft and above - and results in locos which either hobble along most unrealistically at low speeds or else go off like a Guy Fawke's Night rocket. The only way to resolve the issue is either to design the whole loco around a larger and slower-revving motor - as with the dimunitive12/24-powered Black Hawthorn saddle tank reviewed a couple of issues back - or to use a higher gear reduction.

Single-stage gears, at ratios of 50:1 and above, tend to be inefficient and need very accurate meshing. One-piece, double-reduction gearboxes offering ratios in the order of 60:1 and 80:1 are much more efficient and give smoother starting and stoppping. These have been available in kit form for some years now but they tend to be either bulky or noisy or both, nor have they always used the very best components and materials.

Why do we need ratios of this order? Shunting, industrial and narrow-gauge locomotives do not look good or run well when they are teetering along on the brink of a stall - but this is what often happens when you try and run them at a scale speed of 5mph or less, even with compensation and good pick-ups (for gear and motor quality is only part of a complicated formula for good running that includes accurate quartering, decent axlebox design and other factors). Such prototype engines rarely pick up much of a speed anyway and yet the gearing and motoring arrangements of many models equips them out to be at their most efficient and best-behaved at speeds of 40mph and more, which is bizarre.

The discrepancy is not confined to the minnows of the locomotive world. Most of us have layouts whose dimensions and curvature and signalling arrangements would, on the prototype, preclude speeds much in excess of 25mph and perhaps far less. Our small-motored 4-4-0s and suchlike gallop when they should crawl; they stop too abruptly; they accelerate much too quickly, even allowing for the fact that - as I remember things - most steam locomotive drivers went about their work with considerable vim. If you look at logs of actual loco performances you will see that, from a standing start, it took a passenger train about three miles (or six or seven minutes of actual running) to get up to a speed of fifty miles an hour. On a typical layout, the distance from the starting signal to the point of disappearance (into tunnel or fiddle-yard or whatever represents 'the rest of the world') is probably about ten feet, or an actual 250 yards or so. Even with a fire-eater like Sammy Gingell or Bill Sparshatt at the regulator, the train shouldn't be doing more than about 20mph by the time the tail lamp passes from view. And yet we build models that should only really be at home - in terms of efficiency and smoothness - running flat out on large tail-chaser layouts set in open country, where high speeds are far more plausible.

Enter Backwoods Miniatures, punning...
These three new gearbox kits from the punningly named Porter's Cap Productions - a joint venture between Pete McParlin of Backwoods Miniatures and Chris Gibbon of High Level Kits - bring a measure of common sense to the situation. The basic etched units are capable of being adapted to various configurations and ratios, either by simple interchangeability of the gears or by easy modification of the nickel-silver etch. The smallest, the Compacto, draws on Pete's experience in producing bespoke gearboxes for his exquisite Collector's Series of narrow-gauge loco kits. It can be had in 80:1 and 108:1 versions, and the only difference is the worm and first-stage gear. Both can readily be attenuated to two-stage 40:1 or 54:1 configurations where space really is at a premium. The 8mm fixing centres accommodate the recommended Mashima 9/16 open-frame and 12/20 or 12/24 can motors with 1.5mm shafts.

The Compacto is little bigger in size than a conventional 38:1 etched mount and can be used - as indeed can all three of the designs currently in the range - with driving wheels as small as 9mm in diameter. Not wanting to go too far too soon, I assembled my Compacto in the 54:1 version, which enabled me to get a Mashima 12/24 vertically into the firebox of a 4F. With its 5ft 2ins drivers, this is the classic sort of engine for a 38:1 gearset - and nothing wrong with that, in this particular case - but the higher ratio had a significant edge on slow-speed running while the top end was in the region of a scale 40mph.

The mid-range model is the Compresso. This uses exactly the same gears, only more of them (hence the price hike), and is available in 144:1 and 194:1 ratios - the latter can, if required, be built as a 108:1 gearbox instead. Again the components suit 8mm screw holes and a 1.5mm worm. The moulded gears are of the very highest quality - almost too good, in fact, for model locomotives. As the instructions point out, 'Advances in plastics technology have resulted in a range of components intended for far more intensive operation than will ever be encountered in most scale model railway applications'. The mere fact that a gear is made of metal is no guarantee that it's any tougher or more concentric or more accurately cut than its moulded equivalent - if anything, the reverse may well be true.

I fitted the 144:1 version of the Compresso into an Impetus fireless locomotive kit powered by a Mashima 9/16 and found it gave strong, quiet running and exquisite crawl performance, with no hint of any cogging effect. At shunting speeds the armature is whirring around merrily and the possibility of a stall is remote; top speed with 3ft diameter wheels is in the region of a scale 20mph or so, which is more than enough for this locomotive. The first two stages have the gears revolving freely on Loctite-secured fixed shafts. This is a clever piece of structural design, since the shafts rigidly brace the fold-up mount and negate the tendency of some u-shaped nickel-silver mounts to flex under load. Unfortunately I did the classic thing when assembling my sample and Loctited the first-stage gear on solid. Heat from a soldering iron failed to break the bond so I knocked the shaft out with a drift, at which point the gear fell into a box of tools under my bench. When finally disinterred from its hiding place among those hard, sharp objects it was covered in dust and swarf but, once cleaned up, it ran perfectly. This kind of abuse shows how strong these gears are - by contrast, I have dropped brass gears on to the wooden floor of my workroom and ruined them.

Finally we have the Contorto, the mother of all gearboxes. I went through the many 4mm drawings that have been published in the Journal of the Stephenson Locomotive Society - from Sharmanesque curiosities to the Turbomotive and beyond - and found surprisingly few prototypes into which this remarkable articulated gearbox would not fit. The Contorto is, at heart, the same 97:1 design fitted to the High Level Black Hawthorn, but with the reduction stepped up to 108:1. It's also available in an 80:1 version and the reduction ration can, in both cases, be halved if required - this makes sense on main-line locos where we might justify a higher maximum speed. The great glory of the Contorto, though, is the final drive carriage - a separate unit which pivots freely (though it can be soldered solid) on the third-stage gearshaft. This articulation enables it to slide in under fireboxes and cabs, or to double back on itself, or to drop straight down to suit a high-boilered prototype such as the BR Standard classes. The pivoted shaft could, if required, be made to pass through the loco sideframes. This would allow the main part of the gearbox to be removed with the motor still attached and, more interestingly, for the final drive to be installed at the same time as wheeling-up takes place, the rest of the gearbox being added once the quartering has been done and a smooth-running chassis obtained.

The Contorto would, I feel, be particularly useful in many 'difficult' main-line classes where the conventional wisdom is a Portescap RG4 with MJT alternative sides, a configuration that frightens many people - and not just on the grounds of cost. The superbly smooth Mashima 12/24 would be a good choice - I've heard it said from people I respect that, if this Japanese motor had come on the market ten years sooner, the RG4 might never have happened - although the etched mount provides both 8mm and 10mm spacings to allow something more virile to be fitted. 'Test it to destruction' was the word from Chris Gibbon, so I built up my review sample, hung it on to a Mashima 1628 and ran it all day and half the night without any of the gentle easing-in that I usually accord the drive trains I build. The next day I fitted it into a massive North Eastern Railway T1 4-8-0T hump shunter, in place of the totally inadequate (in terms of slow running) D13/40:1 combination with which I'd built it. This is a model engine that does most of its work at maximum power output and exceptionally slow speeds - the prototype was designed to work round the clock in yards such as Gascoigne Wood, Newport and Tyne Dock, pushing rakes of 40-50 loaded wagons up the humps at at a steady 2mph - and so far the drive unit has worked flawlessly. The T1 weighs 560g and has no problem shifting loads of 1.3kg, dead slow, in near-silence and with no sign - touch wood - of any untoward stress or wear. This success has certainly answered a few questions and dispelled any fears I might have had that such a large motor would quickly rip the heart out of small moulded gears, although doubtless some industrial-strength engineer will tell me that really I should be using tungsten-steel gears with a face width of half an inch. For heavy-duty applications such as this, incidentally, the Contorto is available on request with a 2mm bore worm gear.

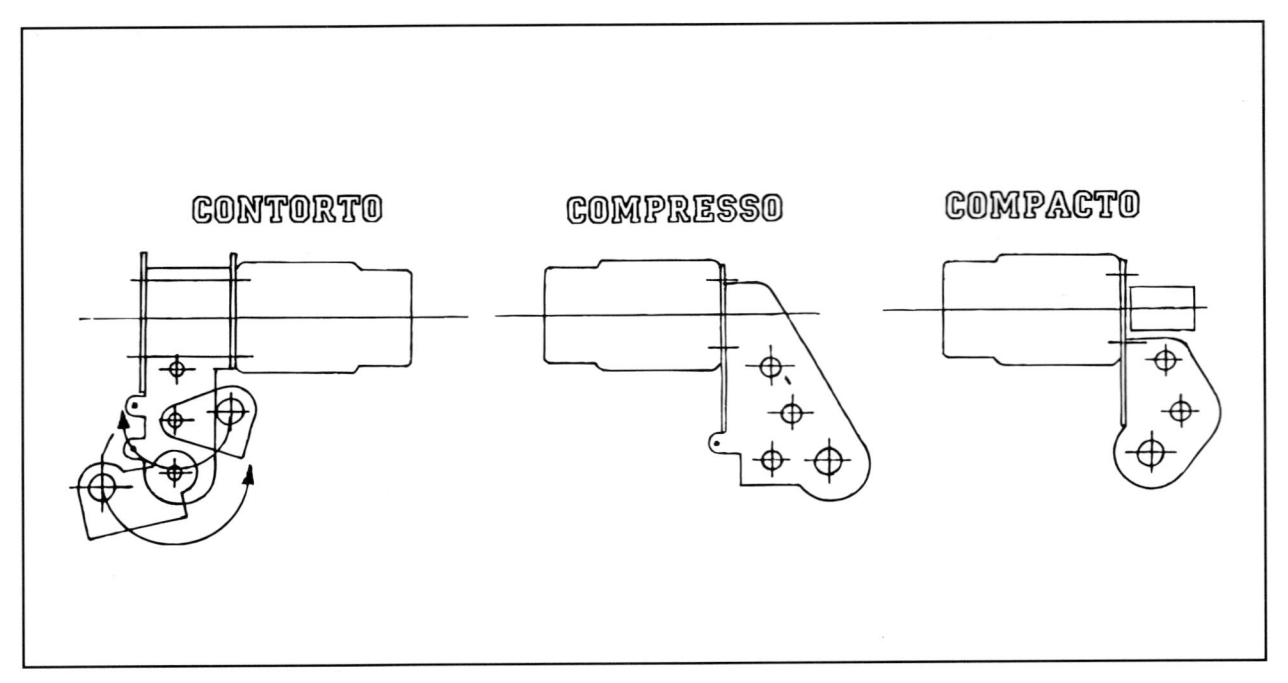

The gearboxes are available by mail order from Backwoods Miniatures at 16 Reivers Gate, Longhorsley, Morpeth, Northumberland NE65 8LA (tel 01670 788 577, fax 01670 788 549). Prices are £12 for the Compacto, £15 for the Compresso and £18 for the Contorto. Please specify gear ratio required (and shaft size when ordering the Contorto) and add 50p towards post and packing. Motors must, of course, be sourced separately but, at under £30 for a drive system of the very highest quality, I believe this represents excellent value for money. To sum up - apart from general-purpose engines with wheel diameters between approximately 4ft 6ins and 5ft 6ins, I no longer feel 38:1 and 40:1 gears are adequate, especially when used with fast-revving small motors. I will continue to use RG4s in some situations with large express engines while my heavy freight locos will be tender-driven via universal joints and reduction gears. For the foreseeable future, though, these three gearboxes will form the basis of my planning for small and medium-sized locomotives - unless Messrs McParlin and Gibbon have something else up their collective sleeve.

Tim Shackleton

Crookes Crafts, Sheffield
Prism RTC Silicon Rubber

Room-temperature curing Silicone Rubber has a number of uses for the modeller, but rarely comes in a convenient form or in small enough quantities for our purposes. Card and art material suppliers Crookes of Sheffield are now supplying a basic RTC silicone rubber sealant made by Prism in 25ml. tubes for £1.70, or 50ml. tubes for £2.50.

The main use I find for the stuff is for sound-deadening on model locos. Mounting drivetrains - complete motor/gearbox units - to the loco with the motor anchored in a blob of silicone sealant is a good way of decoupling the source of vibrations from structures likely to amplify them. This sealant does the job very nicely. The silicone is also useful for retaining lead ballast in loco bodies, and again it can help damp out unwanted resonances - lead encapsulated in silicone rubber has got to be one of the more effective sound-deadeners.

Another use for silicone like this is for making moulds for casting small plaster components - architectural fittings and the like. A friend made a batch of stone sleeper blocks for a plateway model using just such a mould - ordinary spray furniture polish makes a good release agent.   **I.A.R.**

Crookes Crafts, 33 Pickmere Road, Sheffield, S10 1GY
Tel: 01142 668198 Fax: 01142 685339

British Railway Wagons No. 1 OPENS and HOPPERS

*Geoff Gamble, 60pp. 230mm. x 180mm.
Colour covers, 110 B & W pix.
Cheona Publications, £8.95. ISBN 1 900298 01 5
39 The Avenue, Chinnor, Oxon, OX9 4DP*

Well-known railway book dealer and 3mm. modeller Geoff Gamble now has his own imprint, Cheona (ancient form of Chinnor - there's erudite for you!) Publications, and this book of BR wagon photos is is first effort.

The first thing to say is that this is a modeller's source book par excellence. The pictures are all reproduced at postcard size to excellent quality, and contain an absolute wealth of modelling information. The captions, too, are informative, and all-in-all the book covers its subject pretty completely. OK, so there may not be reams of history on lots and diagram numbers (although these are normally quoted in the captions) but, from the modellers point of view, you can see what the thing looks like.

There is a short but pithy introduction on the origins of the fleet, and notes on train class codes and lamp positions, Diagram and Lot Nos., numbering, liveries and the alphabetical TOPS codes. Succinct but useful. The rest of the book is given over the the wagon photos - nearly 100 of them, covering all BR opens from ex-PO minerals to all-steel pallet ale wagons, including that group of late steam-era roller-bearing LWB wagons that no-one ever seems to model . Wonder why not?

My first reaction was that the book was a bit pricey for what it was. But that's the wrong way to look at publications like this. The amount of information contained and the clear and logical presentation - so that you can find the information when you want it - fully justify the price. Look at it another way: How many wagon postcard-size photos at an average 50p. a pop could you buy with £9? about 18.... and you'd still have to do the research to find out what they were.

So, if you're into mutating Red Pandas or playing around with Parksides, this is the book to have. I look forward to future volumes dealing, presumably, with covered wagons and specials.

Iain Rice

Newton Abbot
Rail 150
Model Railway Exhibition

Help the people of Newton Abbot celebrate 150 years as a railway town, with this one-off commemorative event.

- 20 Layouts, including Coldrennick Road 7mm, Mike Sharman 4mm, Ashburton EM, Arcadia 7mm, Ash Branch 3mm, Lydtor 2mm, Bodmin 4mm, Kingswear 2mm, Eggesford 3mm, Charmouth 7mm. NG, Hepton Wharf 4mm, Aberhafren Shed 4mm, Chiseldon MSWJR 4mm, Ulpha Light Railway 4mm, Bishport LMS 4mm.

Extensive trade support; Railwayana and Railway Art - free exhibition at Forde House - free vintage bus link to Newton Abbot station and Forde House.

4 - 5 January 1997
Newton Abbot Racecourse

10.30 - 17.30 Sat. 10.30 - 16.30 Sun.
Tickets £4 on door, OAP/UB40/Under 16 £2.
Advance Bookings £3/£1.50
Details: Roger Bradford, 24 Ashleigh Way, Teignmouth, Devon, TQ14 8QS
Please Make Cheques Payable to: Model Railway Heritage Trust.

Exhibition organised by Model Railway Heritage Trust for Newton Abbot Town Council

Still Available ...

Furness Railway 150
64pp. Many Illustrations
£5.95

Introduction to 4mm. Fine Scale
Edited by Iain Rice
64pp. 87 B & W pix, 28 diagrams £6.95

RailMODEL PO Box 2 Chagford Devon TQ13 8TZ
All titles post free. Access & Mastercard welcome.

Please mention RailMODEL Digest when replying to our advertisers

This time's Good Idea from FourTrack Models
Get a clean finish with FLEX-I-FILE & FLEX PAD from Creations Unlimited

These wonderful burnishing appliances are both available in a choice of four grits

- Course • 150 grit
- Medium • 280 grit
- Fine • 320 grit
- Extra fine • 600 grit

FLEX-PADS are £2-30 each
FLEX-I-FILE starter pack, £6-85
contains U shaped metal handle and
2 each of the course, medium and fine bands.
Packs of replacement bands are available in
Course, Medium, Fine or Extra-fine,
six bands of a single grit size, £1-85 a pack

SAE for list of my kits and other stock items

P&P. UK orders over £20, Free. Other UK £1
Sorry, I can't accept charge or credit cards

FourTrack Models
22 Grange Road, HARROW,
Middx, HA1 2PP
Tel : 0181 863 7338
Exhibitions and Mail Order only

RATIO PLASTIC MODELS
QUALITY KITS FOR THE RAILWAY MODELLER

COLOUR CATALOGUE £3.00
DIRECT FROM THE FACTORY £3.50

AVAILABLE OCTOBER/NOVEMBER

TRACKSIDE OO/4mm
Ref 540 LOCOMOTIVE SERVICING DEPOT
(figures not included)
Price T.B.A

NEW OO/4mm also available
- Ref 543 Hoist from Ref 540
- Ref 436 Security Fencing
- Ref 455 Modern Street Lights

N/2mm
- Ref 455 Retaining Walls

Prices T.B.A.

VISIT OUR FACTORY SHOP
SEE OUR VARIOUS DIORAMAS AND 'N' GAUGE LAYOUT "NETHER STOWEY"
OPEN: Mon-Thurs 8am-5pm Fri 8am-2pm
Closed Weekends and Bank Holidays

RATIO PLASTIC MODELS LTD.,
Hamlyn House, Mardle Way,
BUCKFASTLEIGH, Devon. TQ11 0NS.
Tel: (01364) 642764

Manufacturers of Model Railway kits in N, OO & O Scales since 1950

OUTRAGEOUS!!!

LANCASHIRE
RUSSELL

MANUFACTURER MAKES NARROW GAUGE MODELS THAT ACTUALLY LOOK LIKE THE PROTOTYPE – SHOCK, HORROR!

4mm scale narrow gauge has progressed immeasurably since the days of dubious whitemetal kits on wholly unsuitable RTR N-gauge chassis. Our mould-breaking range of kits treat the whole subject seriously, allowing all the prototypical details which make these locomotives so fascinating to be faithfully modelled.

If this concept appeals to you, why not join the hundreds of modellers who've discovered the joys of Backwoods Miniatures' rapidly expanding range – 35 kits at the last count – covering Ireland's 3ft gauge, together with Welsh, English and Scottish 2ft/2ft 6ins byways.

CHEVALLIER

NO N-GAUGE CHASSIS FOR US, VOWS PROPRIETOR

These kits are perhaps unique in being offered complete with wheels, reduction gearing and Mashima motors. The quality speaks for itself. Details can be obtained in return for an A5 size SAE.

BACKWOODS MINIATURES
SETTING STANDARDS IN 4mm NARROW GAUGE
16 REIVERS GATE, LONGHORSLEY, MORPETH,
NORTHUMBERLAND, ENGLAND, NE65 8LA
Telephone (01670) 788577, Fax: (01670) 788549

Please mention RailMODEL Digest when replying to our advertisers

SHIRE SCENES
4 North Cots, Landscove, Ashburton, Devon TQ13 7LU
Tel/Fax 01803 762498. Post free in UK; Access/Visa/Amex

GWR PASSENGER LUGGAGE VAN K15/K16
KIT LESS WHEELS, PAINT & TRANSFERS. SCREW COUPLINGS, HANDRAILS & END DETAILS ETCHED. DESIGNED FOR EASY CONSTRUCTION WITH FINE DETAIL BUILT IN TO PLEASE THE NOVICE & SATISFY THE EXPERT £49·95

Etched brass made easy. 3 x 1st class stamps for lists

GM&S Godfather Models & Supply
Texelhof 179 2036 KG Haarlem The Netherlands
Tel+Fax (0031) (0) 23 5333106

GM&S levert de Rail Model Digest in Nederland

U hoeft dus niet zonder te zitten als u terug bent uit UK
Wij leveren voorts een groot aantal Engelse modellen en toebehoren van:
Shire/Mashima/spuds/Dapol/Dart/D&S/Slaters/Ratio/Branchlines/Expo/Carr
Springside/Parkside/DJH/Wills/Escap/MMS/GW-tools/ScaleLink/Tower/boeken
tijdschriften/Pressfix/Exactoscale/Dormaplas/Langley/Gibson/MJT/Romford/enz.

More pages have been set aside in this Issue to allow our advertisers to tell you about their products - please support the advertisers that support us!!

If you have anything you want to sell, don't want to pay exorbitant rates but still want to reach a discerning audience, then send for a rate card to:

Helen Chapman,
PO Box 2, Chagford, Devon, TQ13 8TZ
Phone or fax 01647 433611

WHEELTAPPER COACHES
4mm Printed Side Coach Kits

Originally by P.C. Models - "built-in finish" kits with ready-glazed sides, accurately painted and fully lined. Complete kits from £19.50 post paid, including wheels, scale couplings, transfers and interior detail, in a sturdy box which holds the finished model. Coaches from GWR, LMS, LNWR, LNER and LSWR - and components

FREE catalogue - SAE to -
P.O. Box 28, (D) HEATHFIELD, East Sussex TN21 9ZY

New Revised & Updated Loco Kits Available

Revised J21 kit complete with etchings and revised chassis.
New J25 based on the above but with additional fittings
New revised Lancashire & Yorkshire kits for
Barton Wright 0-6-0 and Aspinal 2-4-2T.
HAA Wagon, 11 new L.S.W.R. and 5 new L.N.W.R. coaches now available.
Watch out for coming announcements for a new range of L.M.S. and Southern locos which are in the advanced stage of preparation
Watch this space.

3mm

DART CASTINGS
4mm WHITEMETAL KITS & ACCESSORIES

ENHANCE YOUR LAYOUT WITH MY DETAILING ACCESSORIES

LINESIDE		ROADSIDE	
GROUND FRAME & PHONE CABINET	£2.00	RELIANT 3 WHEEL VAN (1920's to 1960's)	£5.20
POINT LEVERS (4 yard type)	£1.25	POST BOXES (2 pillar, 1 wall, 1 on post)	£2.45
FIRE BUCKETS (6) & STAND	£1.40	DUSTBINS (5 each of two shapes)	£2.40
GWR PLATFORM SEATS (3 wooden)	£4.10	WAR MEMORIAL (for street or churchyard)	£4.10
GWR DOUBLE GROUND SIGNAL	£1.75	SIGNPOSTS (3) (2 styles available)	£3.25
LMS PLATFORM SEATS (3 MR rustic)	£4.10	GRAVESTONES (6 different)	£3.20
GWR MILE POSTS (4, 1 for each 1/4 mile)	£1.40	WELLHEAD WATER PUMPS (3 types)	£3.25
STRETCHER CABINETS (2, for SR stations)	£1.50	HORSETROUGH & DRINKING FOUNTAIN	£2.00
SR GROUND SIGNAL	£1.40	MOTOR SCOOTER & RIDER (1960's era)	£4.00
GWR OIL LAMPS (4, on 5' posts)	£4.10	DUCKS (12 for village pond)	£1.40

DETAILING KIT FOR DAPOL/AIRFIX/MAINLINE GWR AUTOCOACH £15.50.
DO YOU HAVE AN ITEM YOU WANT TO DUPLICATE? IF SO PLEASE SEND FOR DETAILS.
ASSEMBLY WITH SUPERGLUE OR EPOXY. POST FREE IN U.K. SORRY NO CREDIT CARDS.
Mail Order, Exhibitions & Model Shops. Send SAE for list No. 4. Quote: REF. RMD
27 FREMANTLE ROAD, HIGH WYCOMBE, BUCKS HP13 7PQ

EILEEN'S EMPORIUM
55 REEDSDALE GARDENS
GILDERSOME
LEEDS LS27 7JD

SUPPLIERS OF TOOLS & MATERIALS

BRASS, NICKEL SILVER ETC.
ROD SQUARE SHEET WIRE ETC.
TELESCOPIC FINE TUBES
BA SCREWS - NUTS - WASHERS

LARGE SELECTION OF BRASS SECTIONS
GERMAN & U.S.A.

PLASTIC SHEET - STRIP - SECTIONS

SMALL HAND & LATHE TOOLS
GOOD SELECTION OF SMALL DRILLS
CUTTING BROACHES

SWANN MORTON PRODUCTS

PLEASE ENQUIRE
SORRY NO LIST AVAILABLE
TEL/FAX 0113 253 7347
9 - 6 MON - FRI.

C&L FINESCALE

From flexitrack to fishplates, templates to tie-bars, chairs to crossing vees and sleepers to switch blades, we can supply you with high quality components to construct accurate and realistic trackwork in 4mm and 7mm scales.

Some examples from our range :-

		4mm	7mm
Functional running rail chairs - 3 bolt	250 pack	£6.25	£7.45
Functional running rail chairs - 2 bolt	250 pack	£6.25	£7.45
Code 75 bullhead rail, n/silver or steel	10 mtrs	£4.90	
Code 125 bullhead rail, n/silver or steel	10 mtrs		£13.25
Crossing vees n/s or steel - 1:5 to 1:8	each	£3.65	£4.65
Switch blades n/s or steel - A or B type	per pair	£2.95	£5.45
Plastic fishplates	24 pack	£1.70	£2.35
Templates	each	£0.65	£1.25
Flexitrack (OO, EM, P4, O) n/s or steel	metre	£2.75	£4.90
Point kit (plastic components only)	each	£5.40	£7.95

Mail order service available. £1.50 p&p, post free over £30, except for £1.25 additional when rail of flexitrack (minimum 5 mtrs) included in order. For catalogue please send 4 x 1st class stamps plus 9 x 6 SSAE.

Now available: The Tortoise Slow Action Switch Machine - the ultimate point motor! £11.95 each, or £110 for 10.

Why not visit our stand (we attend many of the finescale exhibitions). Our products are also available from Colin Ashby's exhibition stand.

PO Box 45, Harold Hill, Essex, RM3 0DW
Phone and fax : 01708 344063

Please mention RailMODEL Digest when replying to our advertisers

Cherry Paints

Authentic railway pre-grouping, grouping & steam-era B.R. livery paint in 14, 50, 125 & 250ml size tins.

Available from good model shops
or by Mail Order from:
Micrologica Systems Ltd.,
Ronty Brig, Glenquiech, By Forfar,
DD8 3UA.
Send three 1st class stamps for colour cards & list.

Access / VISA

Double O Gauge

Just Toys, Right? - WRONG!
Just for Beginners - WRONG!
Not for Serious Modellers - WRONG!
Some popular misconceptions. In fact,
OO Gauge 4mm Scale can be as finescale as you want!

The Double O Gauge Association

Is an organisation which caters for both new & established OO Modellers. We have new and exciting ideas, and improvements on old themes. Above all, we have fun with our hobby.
Join our growing International Association
Just send an A5 SAE to:
The Double O Gauge Association
PO Box 100, Crawley,
West Sussex. RH10 1YP

TOWNSTREET

MODEL ARCHITECTURE AT ITS FINEST

STONE CAST BUILDING SECTIONS PROVIDE THE REALISM OTHER MATERIALS LACK. SO EASY TO USE & PROBABLY THE BEST YOU CAN BUILD.

JUST ONE EXAMPLE FROM THE RANGE
THE 'OO' BRICK GOODS/ENGINE SHED
EASY TO ASSEMBLE INTERCHANGEABLE SECTIONS ALLOW FULL CHOICE OF DOOR POSITIONS.

Basic shed with end doors £32.50
As illustr. with side door/canopy & office £39.50
Interior Loading Platform Set £6.95

PLEASE ADD £3.00 P&P 'OO'

SEE HOW YOU CAN CREATE REALISTIC STRUCTURES ON YOUR LAYOUT WITH THE TOWNSTREET 'OO' CATALOGUE. 32 PAGES PACKED WITH BUILDING IDEAS, HUNDREDS MORE REALLY USEFUL CASTINGS & FULL COLOUR CENTRE SECTION. ESSENTIAL FOR LAYOUT PLANNING & DEVELOPMENT AT ONLY £2.95 POST FREE OR SEND SAE. FOR 7mm LIST.

**TOWNSTREET, THE OLD SCHOOL
CARNBEE BY ANSTRUTHER
FIFE KY10 2RU**

TEL. ORDERS MON TO FRI 2-8 pm 01333 720226

HENRY WILSON BOOKS

Books, Magazines, Official items on all aspects of prototype and model railways from one of Britain's largest stocks of antiquarian, second-hand & new titles.

Free book search service.

**14 Broomheath Lane, Tarvin,
Chester, CH3 8HB**

Tel/Fax 01829 740693

MasterCard   **POSTAL BUSINESS ONLY**   *VISA*

AMBIS Engineering Division
Moving Modelling closer to the prototype in operation and appearance.

27 Stanhope Gardens, Ilford, Essex IG1 3LQ for Mail Order

Winter 1996: AMBIS have over 100 items listed in their guide. Including lever frames, mechanical interlocking, near scale stretchers, fishbelly rail, simulated tram track parts; parts for wagons, including axleguards in 7mm scale, and the Eastwell Iron Works corrugated iron products.

Late 1996 into 1997
EXPO-South
Watford
Reading GOG
St. Albans
Bletchley GOG

Sample prices:
Long Handled Lever Frame (6/9 etc.)   - from £28.00
Point Actuating Locking Unit (PALM)   - £2.00
Detailed Stretchers for 7mm scale   - from £1.95
or 4mm scale   - from £1.60

More 4mm Wagon Parts Coming Soon

Complete 4th Edition Illustrated Guide
4*1st + 2*2nd Class Stamps

Art Materials by Mail Order
CARD MODELLING SPECIAL

FOAM BOARD
30" x 20" 3mm £3.80 5mm £4.20 10mm £5.00

KNIVES & CUTTERS
OLFA STANLEY SWANN MORTON MAPED

ADHESIVES
EVOSTICK BOSTIK ARALDITE COPYDEX UHU COW GUM 3M

COLOURS & BRUSHES
WINSOR & NEWTON Daler Rowney berol *Karisma*
full ranges available

AIRBRUSHES
De VILBISS, BADGER, AZTEK

This is but a small selection of our vast range, to see it all one of our illustrated catalogues is essential, for a copy just send me an A4 sized SSAE

Enquiries and orders welcome at
CROOKES CRAFTS
33 PICKMERE ROAD SHEFFIELD S10 1GY

tel : 0114 268 6028   fax 0114 268 5339
Postage is £1.00 per order (Heavy items at cost)

FALCON BRASS KITS

159, CLOPHILL ROAD, MAULDEN,
BEDS. MK45 2AF
Fax & Telephone 01525 861187

A SELECTION FROM OUR GWR RANGE

LK26 GWR ARMSTRONG GOODS	£49.75
LK111 GWR/BR 1101 CLASS	£36.75
LK129 GWR/BR 'METRO' TANK	£38.50
LK147 GWR/BR 'COUNTY' CLASS 6-6-0 & HAWKSWORTH TENDER	£55.50
LK148 GWR/BR 'CASTLE' CLASS 4-6-0 & 4,000g TENDER	£57.75
LK221 GWR/BR 38XX 2-8-0 & 3,500g TENDER	£52.50

Copied by many, equalled by none.
(Quote from a customer - Falcon kits are just the right degree of complexity!!!)
Updated catalogue listing 34 tenders, 150 chassis, over 300 locos, 115 wagons and coaches. Only wheels, gears & motors needed for loco kits & wheels for the coach & wagon kts. Locos have turned brass fittings, full nickel silver chassis & valve gear. Spacers for OO/EM/P4 & fine scale hand rail knobs. Pre rolled boilers and other small parts in whitemetal. Having our catalogue could save you from £9 - 40 on prices of other makers similar kits!

Catalogue 50p (stamps) plus long SAE. P&P kits £1.50, fittings 75p

---

Baseboards to Complement your Etched Kits!

Fine Scale excellence in timber or ply to a good exhibition standard. Not suitable for R.T.R. track or children's train sets.
No scenic work or track laying.
Stamped s.a.e. for free leaflets.
Reasonable prices.

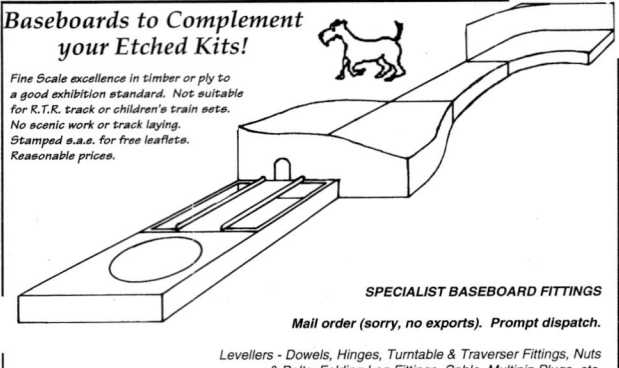

SPECIALIST BASEBOARD FITTINGS

Mail order (sorry, no exports). Prompt dispatch.

Levellers - Dowels, Hinges, Turntable & Traverser Fittings, Nuts & Bolts, Folding Leg Fittings, Cable, Multipin Plugs, etc.

96/97 ILLUSTRATED FITTINGS CATALOGUE £1.50 including p&p

"Red Dog" P.O. Box 1301 Bradwell Milton Keynes MK13 9LA

Please mention RailMODEL Digest when replying to our advertisers

COVE MODELS

The little shop with the big stock and the big service to match.
Large and varied range of all that's best in finescale modelling.

Loco Kits
DJH, S.E. Finecast, Crownline, Alan Gibson, Nucast, Little Engines, Malcolm Mitchell, GEM, Springside, Craftsman, Slater's, J&M.

Coach Kits
Slater's, Alan Gibson, Ian Kirk, Mailcoach, Ratio, Comet, Southern Pride, Shirescenes, Roxey.

Wagon Kits
Parkside, Ratio, Coopercraft, Pocket Money Kits, Chivers, D&S, Colin Ashby, ABS, GW Wagons, Cambrian, Slater's, S.E. Finecast, Red Panda.

Our stock and variety of accessories/parts is huge, many customers ask how we keep track of it, well being only human sometimes we don't but if we haven't got it we will get it, if it is not already on order.

Quality kit building and painting service.

MAIL ORDER no problem. If we can get anything to New Zealand we can manage anywhere nearer.

We are the home of **BLACKSMITH MODELS,** one of the largest ranges of etched kits and parts for 2mm., 3mm., 4mm. & 7mm.
Illustrated Catalogue on this range £1.50 inc. postage

We attend many Exhibitions during the Year and are always more than willing to bring any product along for your collection

44 Cove Road, Farnborough, Hants. GU14 0EN
Tel: 01252 544532   Fax: 01252 376098

VISA, ACCESS, AMERICAN EXPRESS, DINER'S CLUB, SWITCH, DELTA

The Scalefour Society

If you model in 4mm. scale, The Scalefour Society has a great deal to offer you. Those who aspire to create a satisfying and authentic model railway will gain inspiration, companionship, support and information from membership, whether they work in 'OO'. 'EM', 4mm. narrow gauge or, of course, P4

Other benefits of membership include:
Scalefour Stores
Supplying a comprehensive range of products to members with a fast and efficient mail order service. Telephoned orders by Access/Visa are a speciality.
Scalefour News
A high-quality magazine packed with articles of interest to the serious modeller, along with informed comment, letters, reviews, drawings and excellent photographs.
The Area Group System
Exists to keep the membership in touch, and to provide help, encouragement and advice to all those who seek it.
Concessionary admission to Scaleforum
The most prestigious event in the finescale calendar, as well as regional events, such as Scalefour North, Manchester Finescale Exhibition and the AGM Exhibition.

You will find a warm and informal welcome to membership of the Scalefour Society. For a prospectus giving full details send a stamped, self-addressed envelope to our Membership Secretary:

Brian Pearce, 1 Eastcote Road, Pinner, Middlesex, HA5 1DS

For all your railway modelling requirements visit the Ian Allan Bookshops

EVERYTHING FOR ENTHUSIASTS!

LONDON
45/46 Lower Marsh
Waterloo
London SE1 7SG
Tel: 0171 401 2100

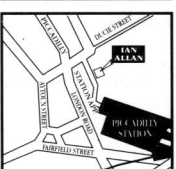

MANCHESTER
Unit 5, Piccadilly Station Approach,
Manchester, M1 2GH
Tel: 0161 237 9840
Fax: 0161 237 9921

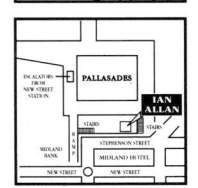

BIRMINGHAM
Unit 84, 47 Stephenson Street,
Birmingham, B2 4DH
Tel: 0121 643 2496
Fax: 0121 643 6855

Each shop stocks or can obtain a comprehensive selection of models from reputable companies such as Bachmann, Lima, Hornby, Graham Farish, Peco, etc. Also available is a wide range of railway, bus, tram, aviation, military, maritime and general interest books plus magazines, models and videos.
Titles/models not in stock may be ordered.

IAN ALLAN Bookshops

RED PANDA
1/76 - 4mm./1ft RAILWAY MODELS

CAT No. RK02
BRITISH RAILWAYS PLYWOOD SHOC-VAN. Dia 1/220, lot. 3224. Also covers lot. 3117 to dia 1/218. Body suitable for lot. 3109.
Three pages of prototype notes and instructions
Available Now .... R.R.P. £4.35

Cat. No. RK01
BRITISH RAILWAYS LOWFIT WAGON. Dia 1/002, lot. 2998.
Eight panel steel sides. 8-shoe clasp vac u/f with Offset Vee's and Lift-Link brake levers. Comes with extra buffer beam (sic) and coupling bar enabling production of Twin Matched pairs. Details of all modifications to earlier lots and Twins, on
Two pages of data .... R.R.P. £3.00

Cat No. RA01
B.R. 10' W.B. UNDERFRAME.
8-shoe clasp brake, vac-fitted with Lift-Link brake levers and Offset Vee hangers. Comes with **Three pages of info and data** listing ALL B.R. wagons originally built using this chassis, along with all known running, lot and diagram nos. Plus details of simple conversion to 9' W.B. (Dia 1/108-109-117 MCV)
R.R.P. £1.50

RED PANDA Railway Models
Do not include wheels, couplings, paint, glue or transfers.

Available direct from:

McKENDRICK MODEL ARTS
PO BOX 3567, LONDON, N1 9HQ

Make cheques/P.O.s payable to **McKENDRICK MODELS**
Add 50p p&p per order. Orders over £10.00 post free
ALL ENQUIRIES : Tel/Autofax 0171 833 0735

ALSO AVAILABLE FROM:
STEAM AGE, 17 Coombe Drove, Steyning BN44 3PW
PARKSIDE DUNDAS, Kirkaldy, Fife KY1 2NL
K & S MODELS, Stevenage, Herts SG1 3AW
THE BOOKING HALL, Charlotte Place, London W1P 1AQ
ENGINE SHED, 749 High Road, Leytonstone, E11 4QS
BEATTIES (Holborn), London WC2
and selected GAUGEMASTER and PARKSIDE DUNDAS stockists.

Please mention RailMODEL Digest when replying to our advertisers

COMET MODELS
QUALITY, ACCURACY AND VARIETY IN 4mm SCALE

SIZE IS NOT EVERYTHING

LCP8 Chassis pack under the Hornby body - £9.45 + p&p

LCP22 Chassis pack under the Dapol body - £9.45 + p&p

Even the smallest of locos can be improved with a COMET Chassis Pack

Catalogue £1.75 inclusive of post.
Post & packing £1.00. Orders over £50 post free
Recent release - chassis packs for D49/1, J72, J39 and LNWR G2
See us at <u>WAKEFIELD</u>, <u>MANCHESTER</u> and <u>WIGAN</u>

105 Mossfield Road, Kings Heath, Birmingham B14 7JE
Tel 0121-443 4000 or 449 5038 (Answerphone)

SHARMAN WHEELS
THE 4mm WHEELS SPECIALIST.

Do you realise that Sharman Wheels have up to five lengths of axle per gauge, have you ever wondered why?

No, it is not because Mike Sharman cannot use a micrometer!!, it is because Mike made the wheels to match the prototype, so if the real thing had a thick boss then so did the scale wheel.

Over the years Sharman Wheels have used various ingenious methods of keying the nylon centres into the steel tyres, these have included filing notches into the rear face, drilling a small hole into the rear face and drilling an indent inside the tyre.

We can announce that in our search to continuously improve the product we have invested in more modern machinery that has allowed us to incorporate a unique keying process within the machine cycle, this locks the tyre on so positively that the only way to remove the tyre is to destroy the centre completely.

We are Steven & Angela Hodgson.

Manufacturers and retailers of :

SHARMAN WHEELS
Glan Henwy, Golan, Garndolbenmaen, Gwynedd. LL51 9YU
Send 2 X 1st class stamps for Catalogue

ROCAR

THE HALT, ROSHVEN, LOCHAILORT, INVERNESS PH38 4NB Tel : 01687 470284

D & S MODELS
NEW INTRODUCTIONS FOR AUTUMN '96 - 4mm.
DS 451 Midland 20 ton 6 wheel Brake Van - £9.90
DS 633 SECR 20 ton 6 wheel Brake Van - £9.90

We are producing two versions of the above design, the Midland type was built over the period from 1886 - 1915 being a larger 20ton-6 wheel version of an existing type, 670 vans built with a further 63 four wheelers. Both types lasted into the BR period and the kit builds either.
From 1898 to 1909 the SECR built 40 almost identical vans, these were later re-built with a double veranda being added to 50 built from 1910 - 14. Again these lasted to the BR period.
All types etched brass with cast whitemetal fittings, fully compensated.

DS 634-SECR-SR-BR 20TON - 6 WHEEL BRAKE - DOUBLE VERANDA.
SR Dia 1558. Please state type required
Price £9.90 each, 90p P&P per order

DS 144 GER/LNER 20TON - 6 WHEEL BRAKE
This is the six-wheel version of the brake van reviewed in the Preview edition of the 'Digest', 50 vans built to this design in 1903 and they ran together with the four wheel version into the BR mid 50s period. Cast body on fully compensated etched underframe.
Price £8.90 each, 90p P&P per order

DS 68 LNER (GC) 15TON VAN - £12.50 (4mm)
This van is now available in **7mm** - Price £32.00 each, £1.50 P&P per order

New Autumn lists
4mm - 80p + SAE, 7mm 40p + SAE
Mail Order Only

D&S MODELS, 46 THE STREET, WALLINGTON, nr BALDOCK, HERTS, SG7 6SW
TEL: 01763 288353

Please mention RailMODEL Digest when replying to our advertisers

HEADSTOCK PUBLICATIONS

NOW AVAILABLE - MY LATEST BOOK
PRIVATE OWNER WAGONS
Published by Oakwood Press
A5 format 88 pages with 83 photographs, almost entirely previously unpublished,
£7.95 + £1.00 p&p (cheques to W. Hudson)
I still have a few copies of
PO Wagons Vols 3&4 £9.95 + £1.00 p&p.

I am able to supply new titles from most publishers and have constantly changing stocks of secondhand and out of print titles and am always keen to purchase such material.
SAE for current list of new and selected secondhand titles. Book Search Service and mail order with pleasure (cheques to W. Hudson)
NORTH MIDLANDS MAIN STOCKIST FOR HAWKSHILL PUBLISHING

Matlock Bath Model Railway Museum
Temple Road, Matlock Bath, Derbys
DE4 3PG
Tel: 01629 580797

The Model Railway Heritage Trust

Modelling Orgy at Missenden Abbey 21-23 March 1997

There has been a regular Railway Modellers' weekend at Missenden for many years. Following research among the attendees on the course, this has been facelifted for 1997, and is now run in conjunction with the Model Railway Heritage Trust.

With the accent firmly on railway modelling, the course will include seminars on key aspects of the hobby, led by leading exponents in their fields, together with practical sessions, and clinics to resolve individual modelling problems. In addition, the course leaders and lecturers will be available throughout the weekend for advice and discussions.

Lecturers and course leaders already booked for the weekend include, among others, Iain Rice, Martin Brent, and Mike Sharman. The course has always proved to be very popular, and early booking is advised.

The Model Railway Heritage Trust is a non-profit making company, limited by guarantee, set up for charitable purposes.
For more information on
The Model Railway Heritage Trust, and the
Modelling Orgy at Missenden Abbey,
please call Pat Barnett on 01923 247851,
or book with the Abbey direct on 01494 890296.

Conserving the Best of the Past for the Future

IMPETUS

7MM NEW RELEASE
LNER Y10
Originally built for the Wisbech & Upwell Tramway these locos also worked at Ipswich, Yarmouth, Kittybrewster (Aberdeen Docks), Edinburgh and Norwich.
It consists of an etched brass body, skirts etc. with rivet detail and dummy boiler with a simple fold up chassis. Castings are included for the tank filler, whistle and buffers, all other detail ie. cab to be added by the purchaser.
7mm £51.00

LNER Y10 NOW AVAILABLE IN 4MM
Similar spec to the 7mm kit but is designed to be powered by a Tenshodo 24.5 motor bogie **£30.95**

7MM NEW RELEASE
FOWLER 0-4-0
A diesel hydraulic industrial shunter introduced in the 60's. An attractive prototype with its cab veranda. The kit has an etched brass body, nickel silver chassis with white metal cast bonnet and fittings.
7mm - £69.95
ALSO AVAILABLE IN 4mm - £46.85

4MM NEW RELEASE
BARCLAY FIRELESS
This is an etched brass and nickel silver kit of the popular 0-4-0 fireless loco that was introduced in the mid 20's for use in gas works, power stations, munitions plants and paper mills etc. Fittings are cast brass and white metal.
4mm £49.85

Further details on the above kits plus others in the range may be obtained by sending 50p in stamps plus sae to the address below. All prices include postage and VAT. Access & Visa taken. Unless specified all kits require motor, gears, wheels, couplings and pick-ups to choice.

IMPETUS
PO BOX 1472 COGGESHALL ESSEX C06 1UQ

SLATER'S PLASTIKARD

SLATER'S (PLASTIKARD) LIMITED
TEMPLE ROAD, MATLOCK BATH,
MATLOCK, DERBYSHIRE. DE4 3PG

★★NEW★★

SOUTHERN RAILWAY - 6 COMPARTMENT BRAKE COMPOSITE
(Diagram 2401)
REFERENCE 7C023P/7C023E*

Kit comprises precision injection moulded components, comprehensive detail castings, easy to assembly etched brake gear, Pullman type corridor connections, internal detail, wheels and couplings, sprung buffers and illustrated easy to follow instructions

7C023P Supplied with injection moulded bogies
£105.75

7C023E Supplied with injection moulded bogies
£110.75

TEL: (01629) 583993
FAX (01629) 580234

Please mention RailMODEL Digest when replying to our advertisers

APPLEBY MODEL ENGINEERING

Leaders in Modern Freight Vehicles

MANY NEW 4MM & 7MM WAGON KITS

In 4mm :- Type II Russell Container
Cawoods Container -- Clam -- Covhop
New HAA & OBA Underframes.
BP green & yellow and Lever Bros. paint.
Surprise new lines.

Also available -
Our range of 4mm & 7mm wagon, underframe, and conversion kits.
English & Continental buffers including, in 4mm scale, type 60 loco.buffers, bogies and accessories.
Matched paint for our kits includes E.W.& S (Wisconsin) paints maroon & gold, Aircraft Blue for Mainline vehicles and Albright & Wilson green.
An increasing range of transfers.

For copies of our 4mm & 7mm catalogues please send 75p in stamps & a stamped, self addressed envelope (A5)
PO Box 104, Worcester, WR5 2YZ,
Tel/Fax. 01905 351952

SOUTH EASTERN FINECAST

NEW Loco Kit F167
S.E.C.R./S.R./B.R. Wainwright 'P' Class Tank £62.95
NEW Chassis Kit FC167
S.E.C.R./S.R./B.R. Wainwright 'P' Class Tank (will fit old kit) £22.95
NEW Loco Kit F183
S.R./U.S.A. Dock Tank
(inc Special Romford RP 25 wheels & axles £73.50

**4mm LOCOMOTIVE KITS
1/32 STEAM TRACTION ENGINE KITS
1/24 and 1/43 CAR KITS
WILLS SCENIC SERIES PLASTIC KITS
FLUSHGLAZE WINDOWS**

**GLENN HOUSE, HARTFIELD RD, FOREST ROW,
EAST SUSSEX, RH18 5DZ**

Telephone 01342 824711 Fax 01342 822270

NEW BOOKS from *CHALLENGER PUBLICATIONS*

LNER Locomotive ALLOCATIONS
1st January 1923 (the First Day)
compiled by W.B.Yeadon.

This superb new paperback lists where each individual locomotive, inherited by the LNER, was 'shedded' on 1st January 1923 (the First Day). Laid out in pre-group company, and class order, with the shed allocation alongside, the book then goes on to list each engine shed with its allocation in class order. Although no 'official' locomotive allocation record was ever made available for the newly formed LNER, Willie Yeadon has managed over the years to piece together what would have been the situation on the first day of Grouping. Working both forwards, from lists issued in pre-grouping days, and backwards from similar material dating from the 1930s, the compiler with assistance from numerous sources has created a unique and fascinating record. Illustrated with 'period' views, this book is a real 'eye-opener' as to what was where at the time under review. A must for any follower of LNER locomotive history, this record will be of invaluable use also to those whose leanings are to any of the companies which became part of the LNER.
ISBN 1-899624-19-8. Paperback, 48 pages. Price **£6.95**.

Available from all good book shops or direct from:
CHALLENGER PUBLICATIONS, 15 Lovers Lane, Grasscroft, Oldham, OL4 4DP. *Please add 10% postage under £30.*

From Miles Platting to Diggle via Ashton.
by Jeff Wells

The main line from Miles Platting to Diggle formed the primary LNWR route between Lancashire and Yorkshire. With its stiff grades and difficult terrain, the line passed through two towns, Ashton-under-Lyne and Stalybridge, on its climb to the Pennine 'barrier'. Jeff Wells looks at the route in its heyday when 'double-heading' of heavy passenger trains was the norm. Also surveyed are the numerous branches and industrial lines which bisected the line throughout its length. Numerous illustrations compliment the readable and informative text.
ISBN 1-899624-18-X. Paperback 96 pages. Price **£12.95**.

YEADON'S REGISTER OF LNER LOCOMOTIVES
Volume 11. Gresley J39 Class
by W.B.Yeadon

This, the eleventh book in the series charts the life of that veritable 'workhorse' of the 0-6-0s, the J39. With 289 locomotives in the class and production spanning a fifteen year period, there were bound to be many detail differences amongst these 'standard' 0-6-0 engines. All is revealed by the foremost authority on LNER locomotives.
ISBN 1-899624-16-3. 96 pages. Casebound, laminated. Price **£16.95**.